China's Green Consensus

Despite contrasting approaches, democratic and authoritarian governments all underline the fact that environmental protection is crucial and inevitable—and China's enthusiasm in stepping up its efforts to protect the environment has not gone unnoticed. This book highlights how the consensual orchestration of sustainability in China's biggest city, Shanghai, affects non-state actors' ways of perceiving, acting, and organizing around environmental issues.

China's Green Consensus examines grassroots realities as they intersect with events of everyday life, offering insights into areas that far transcend debates over coercive forms of environmentalism and exploring the "soft" and "green" facets of President Xi Jinping's authoritarian approach to governance. The importance of environmental protection in people's lives serves as a lens to analyze and understand authoritarian adaptations to environmental global phenomena. Arantes highlights how, through mobilization and (de)politicization, a "green" consensus leads to the displacement of state responsibilities and the cultivation of civil society in its own image. In so doing, she opens up new ways of thinking about the complexities of environmental governance, consensus politics, subject making, and citizenship in authoritarian contexts.

This book will be of interest to scholars and students of Chinese society and politics, environmental politics, political ecology, international relations, and urbanization in Asia, as well as all others interested in the rising appeal of authoritarianism around the globe.

Virginie Arantes is a Wiener-Anspach Postdoctoral Fellow at the University of Oxford, England. She researches in the areas of environmental politics, governance and ideologies, state–society relations, and urban life.

Routledge Contemporary Asian Societies

Routledge Contemporary Asian Societies provides an original and distinctive contribution to current debates on evolutions shaping societies, cultures, politics and media across North and South East Asia. It is interdisciplinary in its approach and the editors welcome proposals across the social sciences and humanities; from political, social, cultural and economic studies to gender, media, literature, anthropology, philosophy and religion.

Series Editors: Vanessa Frangville and Frederik Ponjaert, Research Centre on East Asia (EASt), Université libre de Bruxelles, Brussels, Belgium

China's Youth Culture and Collective Spaces
Creativity, Sociality, Identity and Resistance
Edited by Vanessa Frangville and Gwennaël Gaffric

History, Memory and Territorial Cults in the Highlands of Laos
The Past Inside the Present
Pierre Petit

China-Latin America and the Caribbean
Assessment and Outlook
Thierry Kellner and Sophie Wintgens

China's Green Consensus
Participation, Co-optation and Legitimation
Virginie Arantes

China's Green Consensus

Participation, Co-optation, and Legitimation

Virginie Arantes

Routledge
Taylor & Francis Group
LONDON AND NEW YORK

First published 2023
by Routledge
4 Park Square, Milton Park, Abingdon, Oxon OX14 4RN

and by Routledge
605 Third Avenue, New York, NY 10158

Routledge is an imprint of the Taylor & Francis Group, an informa business

British Library Cataloguing-in-Publication Data
A catalogue record for this book is available from the British Library

Library of Congress Cataloging-in-Publication Data
A catalog record has been requested for this book

ISBN: 978-1-032-13881-7 (hbk)
ISBN: 978-1-032-13883-1 (pbk)
ISBN: 978-1-003-23132-5 (ebk)

DOI: 10.4324/9781003231325

Typeset in Times New Roman
by Taylor & Francis Books

Contents

Illustrations

Figures

Table

Box

Acknowledgements

Research for this project was funded by a Fund for Scientific Research (F.R.S.–FNRS) Research Grant under the auspices of the Université libre de Bruxelles, a Wallonia-Brussels International Excellence Grant, and a Fieldwork Research Grant from the Université libre de Bruxelles. I am extremely grateful to these institutions, as well as support received from the Wiener-Anspach Foundation and Wolfson College in the latest stages of the writing.

The journey towards completing this book is unquestionably and infinitely indebted to many individuals I had the chance to cross paths with during its completion. Most of all, I wish to express my sincere gratitude to my dissertation supervisors, Professor Vanessa Frangville and Professor Thierry Kellner. Thank you both for your unconditional trust, support, and encouragement. I have been extremely lucky to have been taken under their wings when most needed. Without both their guidance, this work would be infinitely different, certainly worse, and probably non-existent. More than an intellectual journey, Thierry's contagious passion for knowledge and Vanessa's ability to make the impossible possible will keep on pushing me forward. I also thank Dr. Luca Tomini for his numerous inputs and priceless advice. Your generosity has been one of the valuable contributions to this book.

My gratitude is also to my colleagues in China. I am deeply thankful for their warm welcome. The fieldwork wouldn't have been so fruitful if not for their extreme kindness and support. The other most obvious people to thank are all the informants, activists, and scholars I have met in China and beyond. Unfortunately, I cannot cite their names here, but I am obviously indebted to them for accepting to speak with me and share their invaluable knowledge.

The list of colleagues and friends to whom I owe debts of gratitude is a lengthy one. I must start by taking Lisa Richaud: I greatly benefited from her suggestions and our rich conversations (and I still do). Next, all my colleagues from CEVIPOL, Emilien Paulis, Marco Ognibene, Arthur Borriello, Lara Querton, Fanny Vrydagh, Fanny Sbaraglia, Suzan Gibril, Robin Lebrun, Ekaterina Gloriozova, David Talukder, and, of course, Leslie Grietens, who have contributed to making this writing journey such an enjoyable experience. My dear colleagues from EASt, Clémentine Léonard, Coraline Jortay, Hua Bin, Van Minh Nguyen, Nolwenn

Salmon, Flora Lichaa, Sonemany Nigole, and Frederik Ponjaert, many of whom have become friends for life.

Finally, strengths come in many forms, and mine unquestionably from my family and their constant and unconditional support. Merci Maman, for your encouragement, your comments, your corrections, your precious advice and what you help me to be. Obrigada Papa for believing in me. I dedicate this book to you. Thank you, Elodie, just for being there for me.

My most profound debt, however, and words cannot express how grateful I am, goes to Vitor Lemos, my life partner, lover, friend, critic, and best supporter. Thank you, not only for following me around the world, but for (almost) always finding the right word.

1 Introduction: creating a "common green vision"

Introduction

In March 2006, farmers from a small village on the Jinsha River, in southwestern China, Yunnan Province, kidnapped seven hydropower surveyors. Their hope was to stop the construction of the Tiger Leaping Gorge (*hu tiao xia* 虎跳峡) Dam, in one of the world's deepest canyons. The next day, a local official, the deputy country head, came and helped the leader of the group of seven to escape. Trust had clearly broken down and, as the news spread, 10,000 farmers formed a human wall around the government office of Deqin County (德钦县). Night arrived, and it grew cold, but the protesters stayed gathered there. The demonstration ended two days later (without violence), as the farmers retreated on hearing the announcement that the construction of the dam would no longer go ahead. The campaign became one of the biggest success stories of China's "green awakening" (Kang and Hong 2010) and, as Liu Jianqiang 刘鉴强, Chinese investigative reporter and environmental activist, put it, was "a nail stuck in the companies' throat" (Geall 2013, 206).

This is one example among many that illustrates the bottom-up public demonstrations—against coal-fired plants, chemical plants, waste incinerators or dams—which arose all around the country at the beginning of the 2000s. Among this wave of protests, the demonstrations against air pollution were particularly powerful. Elizabeth Brunner (2019b) writes of one such protest, describing a silent, masked demonstration in Chengdu in 2016 when the "smog was so thick residents could not see the spotlight at intersections". Because of their dynamism and strength, the Chinese citizens' anger at air pollution gave rise to a breath of optimism, leading many to believe that now, following the collapse of the Soviet Union, China would take a democratic turn. But as Latour (2017) and other scholars argue, Fukuyama's notorious "End of History" thesis seems long gone. The response of China's leadership to its enormous social and environmental stresses—while sticking to its political model—attest to Fukuyama's failed predictions (as do recent historical events such as Brexit or the election of Trump). Xi Jinping 习近平, General Secretary of the Chinese Communist Party (CCP) and Chairman of the

DOI: 10.4324/9781003231325-1

Central Military Commission since 2013, continues to defy all hopes of a shift towards Western-style democratic liberalism.

Granted, it is not surprising that some commentators doubted the capacity of China's one-party system to respond to its environmental situation and attendant social consequences including citizen discontent (Bruun 2017), the growth of social media (Brunner 2019a) and, above all, the rising number of environmental organizations and activists since the beginning of the 2000s, which came to be termed an "associational revolution" (Tai 2015), and "a boom in associational life" (Salamon et al. 1999). Although the aforementioned Tiger Leaping Gorge incident played a major role in the cancellation of the dam's construction, the movement's success stemmed from the tenacity of what I term "Social Good Organizations" (SGO) in educating, informing and coordinating the actors together throughout the two-year campaign (Geall 2013; Shapiro 2013; Arantes 2016).

Normally referred to as Non-Governmental Organizations (NGOs), or Civil Society Organizations (CSOs) in the literature, the terms "non-governmental" and "civil society" both describe what the organizations are not rather than what they are. They encapsulate strong preconditions about the political reality of Western societies. Rather than delve into a contested debate on the pros and cons of this disputed terminology (Perry 2009; Gauss 2016), or discussing the obvious distinctions between the state, the market and civil society, this study advances the concept of SGO to illustrate the variety of organizations I encountered in the field, thus side-stepping ambiguities related to the "right" definition of an NGO.

It is important to highlight that designing a clear-cut typology on SGOs in China is an extremely complex task. As I will explain further, some organizations develop multiple identities and continuously adapt and reshape themselves to respond to new contexts. Several of the organizations I encountered in Shanghai, for instance, developed their activities through a large variety of institutional logics, for example, as a formal NGO in Hong Kong, as a private enterprise in Mainland China, or sometimes even both.

Suppressed during the Mao era and driven "out of existence" in the mid-1990s following the Tiananmen protests (C. Hsu 2010), many organizations developed in a grey zone, either by registering as for-profit companies with the Ministry of Industry and Commerce or by not registering at all (Saich 2000; Ma 2004; Shieh and Schwartz 2009; Geall 2013). As I explore in this volume, despite having varying degrees of autonomy and different organizational structures, the organizations I observed in the field have two common features: (1) they start out from the grassroots level, and (2) they share the mutual goal of pursuing a "general" public good, and positively impacting society or the environment. The concept of SGO is used to avoid "restrictive" and "contested" notions (e.g., NGO, GONGO) (Najam 1996) and, therefore, better capture the social, political and legal environment of the organizations in question.

The growing number of such organizations, and also Chinese citizens' dissatisfaction *vis-à-vis* the environmental catastrophe, have led many scholars to investigate the contentious environmental movement (Brettell 2003; Lin and Ross 2005; Stalley and Yang 2006; Xie and Ho 2008; Xie 2009; Spires et al. 2014; Reese 2015). A growing number of SGOs have come to symbolize the struggle for more space for civil society. Although existing scholarship has made it clear that organizations faced limitations at multiple levels (e.g., constraining national legislation, lack of legitimacy among the Chinese society) (J. Hsu and Hasmath 2015), the organizations can also be seen as proof that Chinese civil society was "bouncing back from the crushing catastrophe of the Mao era" (Geall 2013, 4).

A growing number of SGOs and environmental movements have been giving form to a more "pluralistic" civil society (van Rooij et al. 2016). Xie (2011), advancing similar thoughts, contended that by expanding their "sphere of conflict" and derailing the construction of dams or chemical plants, SGOs defied the conventional image of political and civic life in an authoritarian context. Drawing on mainstream (and hegemonic) theoretical models (mostly Western American), scholars and observers expected that China's alarming environmental issues would reframe state–society relations. Yet, the story turned out to be very different: environmental movements didn't challenge the Communist Party of China (CPC); on the contrary, their prospects of growth were thwarted (Kostka and Zhang 2018).

In a development unforeseen by most China commentators, Chinese leaders embraced, rather than avoided, environmental concerns. Xi Jinping's determinacy to adopt a "green" vocabulary, as epitomized by the thought of "ecological civilization", points to this new paradigm: the environmental question is not to be taken as a thorny issue but as a comfortable crutch for the party-state to advance its legitimacy nationally and abroad. Against all the odds, and despite being the world's worst polluter, the CCP is taking a "green" turn. Taking advantage of the Trump administration's drop back on environmental policies, symbolized by the United States' withdrawal from the Paris Agreement in 2017, China has since then been on the road to leadership in global environmental governance. Environmental movements, depicted by eminent scholars as key actors for social and political change (Doherty and De Geus 1996; Giugni and Grasso 2015; Della Porta and Diani 2020), add to the long list of failed phenomena expected to weaken China's authoritarian regime. *China's Green Consensus* traces this connection between authoritarian resilience and the domain of environmental politics.

The information presented here will show how anxiety around environmental issues is giving the CCP new opportunities to develop, and therefore instrumentalize utilitarian discourses on nature, such as the concept of "ecological civilization". Constructed around the idea that the CCP is the key to solving the basic problems of pollution and environmental degradation, "green" mechanisms are increasingly colonizing political space regarded as a site for contention (Swyngedouw 2019, xv). This, consequently, contributes to depoliticizing—or placing outside the political—authoritarian modes of governance and repressive forms of control. This book therefore argues that, by embracing environmental stewardship, the CCP obstructs opportunities for contentious participation. Taking inspiration from a debate over coercive forms of environmentalism at the macro-level, the book traces grassroots realities as they interact with events of everyday life across the "sustainable city". Notably, it explores how the consensual orchestration of the city changes local actors' ways of perceiving, acting, and organizing around environmental issues. As such, the book attempts to contribute to three ongoing debates in environmental politics: (1) environmental governance in authoritarian regimes, (2) consensus politics, and (3) subject-making and citizenship in the urban context.

Consensual modes of governance

This book explicitly engages with the pressing question of how the CCP is regaining control over the environmental sphere and, by extension, reinforcing its authoritarian legitimacy. Over the past few decades, scholarship has recognized environmental movements for their role in fostering change across the European and Asian continents (Khrushchev and Gorbachev 1993; Brettell 2003; Ho 2007; Carmin and Fagan 2010; Sima 2011; Geall 2013; Xie 2015). Thus, the increasing number of "mass protests" due to the party-state inaction led observers to expect dramatic changes. Yet, as mentioned above, the response of Chinese leaders to environmental degradation cut through the promises of the expected co-evolution between democracy and environmentalism.

Although environmentalism (in itself) cannot be relied on as the sole factor behind democratization, scholars have recognized the cause as an important component of the changes taking place in the politics and society of transitional democracies (Hicks 1996). Disregard for the environment, according to the scholarship, would lead to an accumulation of social unrest (Economy 2010; Boyd 2013; Zhong and Hwang 2015), boosting the CCP's decay (Pei 2013). This explains why rising social discontent at the environmental situation—as illustrated by the viral online responses to Chai Jing's movie *Under the Dome* (2015)—elevated environmentalism, at least for a time, as a sign of the maturation of China's civil society (S. Wang and He 2004; ; Ho 2007; Sima 2011; J. Y. J. Hsu 2014; Tai 2015).

Environmental problems were undeniably severe in recent years, and the authorities regarded the increasing number of protests as undermining social stability in big cities such as Shanghai, Guangdong, or Kunming. Peaceful protests and riots were on the rise and gradually more urban-based, thus posing additional difficulties to China's leaders as they became less "quiet" (Lora-Wainwright 2017; Johnson et al. 2018). As empirical findings by Zhong and Hwang (2015)[1] or Steinhardt (2019) uncovered, Chinese citizens' awareness was growing fast. For Beeson (2019, 166), this shows that economic growth, and the implicit political bargain between the CCP and the wider population on which it is based, and which has been the foundation of social stability, are being undermined by environmental problems.

Given China's catastrophic environmental situation, and the presupposed incapacity of authoritarian regimes to tackle the problem—although countries such as Singapore increasingly defy such arguments (Han 2017)—the prospects for social and political stability have been ambiguous since the 2000s. The rise of protests and environmental organizations would lead, some argued (T. C. Lin 2007; Lee and Ho 2014; Fengshi Wu and Wen 2014; N. W. M. Wong 2016), to greater political pluralism and, thus, increasingly defy the legitimacy of China's one-party system. Yet since Xi Jinping became President, the CCP's way of organizing around environmental issues has changed. The concept of "ecological civilization", although not new, advances the development of a "good governance" system, whereby participation and disagreement are allowed if they don't defy the party-state's legitimacy. Because Xi Jinping advances himself as a "green" leader, and because the concepts of "sustainability" or "protecting the environment" are so devoid of properly political content, it becomes arduous to disagree with him (Swyngedouw in Krueger and Gibbs 2007). As we shall see later in the book, the CCP has learned to take advantage of the public consensus about the need to be more environmentally sustainable.

This new scheme, I argue, is particularly being used in urban spaces to suppress "antagonism and agonism" (Allmendinger and Haughton 2012) and establish a new form of "green" consensus. Designed around "sustainable" practices and goals, China's "ecological civilization" model sediments a new hegemonic order while reducing the space for contestation. This model particularly calls for "expert" governing and "efficient" policymaking and disavowing disagreement over alternative socio-ecological futures.

Environmental crisis as the new "target"

We typically understand authoritarianism as a style of governance that concentrates power, minimizes political pluralism, and represses civil society. According to Nick Buxton (2017), authoritarian leaders often justify repression in the name of confronting a supposed "enemy". This book will claim that the *environmental crisis* is the new target of China's regime. Of course, Xi's "war on corruption", "fight against terrorism", or campaigns against "Western values" often pinpoint various actors as "enemies" of the party and the state. But none of these "enemies" are as dramatic and imperative for the legitimacy of the party as the *environmental crisis*. Why? Because Xi and his team have learned to capitalize on the "consensual scripting" (Swyngedouw 2011, 77) of the need to address the environmental catastrophe. Through deepening and sophisticated imaginaries, arguments, and cooperative strategies, President Xi advances "ecological civilization" as the sole means to fight against the so-called "climate urgency".

The construction of a Chinese "green" governance model includes several elements that are worth delving into. First, they reveal how, in order to meet ecological goals, it is necessary to follow the CCP's central authority and leadership. That is to say, environmental protection must be grounded in the authoritarian one-party political system (Hansen and Liu 2018). Second, the model calls for expanded public participation in social governance.[2] While these two elements seem contradictory in an authoritarian context, we shall see how this actually allows Xi's government to impinge upon the new roles of urban citizenship that perpetuate the model's resilience. Exploring China's authoritarian regime survival strategies to deal with the climate, *China's Green Consensus* thus builds on existing works (such as Brain and Pál 2018) that question whether environmentalism can fuel authoritarianism.

Environmental authoritarianism

To challenge the dominant conceptions of the relationship between the environment, community participation, and democratization, I use the concept of *environmental authoritarianism*, which I borrow from Mark Beeson (2010, 2016, 2018). The concept dates back to the decline of socialism in the 1980s (Shabar 2015) and has been used to analyze contexts as diverse as the Soviet Union, Iran or Egypt (Ophuls 1973; Ophuls and Boyan 1992; Doyle and Simpson 2007; Sowers 2007; Wells 2007; Han 2015). Yet, it is Beeson's work that inspired a flourishing literature on China's top-down approach to sustainability (Eaton and Kostka 2014; Moore 2014; Kostka and Zhang 2018; Y. Li and Shapiro 2020). Going beyond Beeson's theoretical premises, this book, with its ethnographic approach to the grassroots, asserts that the CCP's recent embracing of environmental challenges leads to the construction of new structures of community organization in which the urban citizen becomes the agent of change. By displacing state responsibilities and promoting individual norms of "everyone is responsible" (*ren ren you ze* 人人有责), non-state organizations—once viewed as key actors of pluralization—become proxies of Xi's new "green" turn.

The empirical evidence presented here challenges a set of presumptions (mostly from the Western world) by showing that environmental issues and a rising number of SGOs are not conducive to democracy or, at least, they do not trigger a weakening of China's one-party rule. By assessing the outcomes of CCP's new "green" strategies on the ground—(re)centralizing environmental governance efforts; creating a consensus around environmental protection ("ecological civilization"), and institutionalizing grassroots movements—this book seeks to expose the drive that has led the attitude of China's leadership on environmental issues to evolve from one of avoidance to one of instrumentalization. It particularly stresses that an authoritarian regime can engage in participative approaches under a green banner to promote fresh forms of environmental citizenship and, ultimately, create a governance system that strengthens its resilience capacity.

Defying dominant discussions of environmentalism (Dryzek 1987; Eckersley 1992) that concentrate on the relationship between democracy and environmentalism, this book stresses that environmental protection can become a playground for authoritarian regimes to pursue ulterior motives and goals. It approaches this topic with the following particular questions in mind: Should scholarship consider environmentalism an additional dimension that reinforces (rather than weakens) authoritarianism? How are public participation and citizenship being shaped in China's urban spaces? What is the role of civil society in today's national and local governance system? Is pluralization emerging from the grassroots? And, if so, how and why does it do so?

Looking at authoritarianism through the city

The research described above aims to assess the nurturing of consensual narratives of "sustainable urbanism" in Shanghai. The book's ethnographically grounded approach provides unique details of how China's "green" consensus is colonizing the urban sphere. My observations here are directed mostly towards understanding how China's *environmental authoritarian* approach to governance has emerged as a result of consensual recognition of the necessity for sustainable urban development. Urbanization has long been recognized as a crucial component of states (Childe 1959) and designed to serve their interests (Lefebvre 2000). I particularly take inspiration here from Michel de Certeau's (1984) investigations into the realm of routine practices to argue that the state "strategies"—to place consensus at the heart of state-society relations and, thus, shape new forms of citizenship—are operationalized through the experience of urbanization, on the one hand, and the work of SGOs, on the other. Chen and Lees (2018) state that urbanization is a good entry point for understanding authoritarian environmental approaches to governance. Notably, how the urban is used to celebrate ecology as the next historic step of the development of civilization—after agricultural and industrial civilization—and influence the way the population perceives their role in society (Y. Li and Shapiro 2020).

Yet, before delving more deeply into the epistemological approach of the book, it is worth asserting that my approach has been carefully constructed in relation to the constraints I faced in the field. When I began the research on which this book is based, around 2014, Xi Jinping had recently taken the reins of the Communist party. At that time, nobody was certain about the changes that China's new leader would bring, but the reinforcement of authoritarianism rapidly became a certainty (Simpson 2016). Following his "election", President Xi grasped all the levers of power in the party and the state—including the military and police—becoming the most powerful Chinese leader in decades. Xi now controls all aspects of economic, political, cultural, social, and military reform (Lovell 2020), alongside the internet and information security.[3] Another important detail is how in 2018, Xi cleared the way to stay in power—indefinitely—with the abolishment of term limits on the presidency. As if that wasn't enough (and also of great relevance for this book), since coming to power Xi has made efforts to counter social unrest and popular movements (O'Brien and Deng 2017; Elfstrom 2021). Crackdowns have resulted in the disappearance of tens of thousands of people. As Aaron Sarin (2020) asserts, Xi has allowed his secret police to kidnap, torture, or detain anyone advancing a heretical point of view. No example better portrays how far Xi will go as the Uyghur concentration camps, while Professor Sun Wenguang's arrest during a live interview with *Voice of America* is also revealing of Xi's coercive methods.[4]

Criticizing and grasping the consequences of these coercive mechanisms is crucial. But the more time I spend in the field, the more I feel the urgency to explore the "soft" and "green" facets of Xi's authoritarian approach to governance. How is the party strengthening his presence in people's daily lives? As I explain in the rest of the book, before I arrived in Shanghai in 2016, the government passed two laws that affected China's SGOs' perspectives for growth. The Charity Law (*cishan fa* 慈善法), for instance, was issued on 28 April 2016, but only came into effect on January first of the following year. The timing here is critical, for as I conducted most of my fieldwork from April 2016 to July 2017, environmental organizations and activists were unaware of how the situation would develop, and were on the defensive.

Thus, wandering around the city turned out to be an essential tool to allow me to respond to the aforementioned questions. The data I gathered during semi-structured interviews was thus complemented by practices of urban everyday life, such as informal conversations with citizens, guards of gated community, garbage collectors, or university friends. I carefully kept all these observations and thoughts in field notes. Documents such as photographs, artefacts, books, reports, maps, PowerPoint presentations, participation cards and so on were collected and analyzed. I also took part in daily WeChat forums conducted by the organizations and activists' understudy. Above all, it was this emic approach to the daily life of the different groups under analysis that enabled me to circumvent the various difficulties attendant to doing fieldwork in an increasingly authoritarian environment.

Combining macro and micro perspectives, the book questions how the greening of the party and the state changes the basic structure of urban organization and governance. The importance of environmental protection in people's lives serves as an analytic lens to reflect on authoritarian adaptations on the environmental global phenomena. How do the changing structures of Shanghai, as an urban space, change and redefine the relationships of people, organizations, and their perception of sustainability and care for the environment? Do these micro-practices in the urban context function as "disciplinary institutions" (Foucault 1975)?

Tracing the spread of sustainable practices

When deconstructing on the "urban fabric" (Lefebvre 1970), be it in America, Africa, Europe, or Asia, scholars tend to agree on two points: (1) it increasingly falls into neoliberal practices (Brenner and Theodore 2002; Pinson and Journel 2016), and (2) it helps political leaders to reach their goals (Tuan

1989). Yet, it has been more difficult to reach a consensus on whether urban processes can be understood through a "planetary" lens, or whether a global pattern obstructs local unique practices (Simone 2014; Schindler 2017; Luger 2018). Such debates have also emerged in the Chinese context. While acknowledging that "the urban" has materialized as the most significant ideological realm in contemporary China, Tim Oakes (2019) stresses that it needs to be theorized in ways that avoid scalar and institutional dichotomies. Kit Ping Wong's (2019) empirical findings in Dongguan (东莞) also reinforce this need to overcome the dualistic approach to properly address the political specificities of China's urbanization processes. Moreover, a consensus is growing on the fact that although the Chinese experience is unique and different from other developing countries' experiences, it can contribute to our understanding of the urban (Hamnett 2020; Fulong Wu 2020).

The research in this book inscribes itself in this conversation by recognizing the need to avoid binary categories (west/east or liberal/illiberal) and acknowledging that urbanization combines global ideas and site-specific socio-cultural factors (Luger 2016). This work is inspired by many discussions and debates on China's great urban transformation among scholars such as Fulong Wu (2010, 2016),[5] Choon-Piew Pow (2009), George C. S. Lin (2007; He and Lin 2015), Lin Ye (2017), Martin de Jong (H. Li and de Jong 2017; de Jong et al. 2013) or Dorothy J. Solinger (1999). I will particularly focus on the complexities and opacities of subject-making in Shanghai's increasingly "green" urban landscape, considering questions such as: How do the complexities and ambiguities of sustainable urbanism shape our everyday lives and inner-self and, ultimately, contribute to *environmental authoritarianism*?

The book argues that Shanghai's "green" transformations and attendant practices of participatory governance are constructed to consolidate the resilience of the CCP while representing this process as a pragmatic response to an urgent need for change. The consensus is reached by various forms of depoliticization that all place environmental issues outside the field of public debate and discourage alternative practices, thus erasing spaces for contestation. These processes are particularly present in Shanghai as the Metropole aspires to put China at the top of the list of the world's most sustainable cities.

In this light, the book follows a growing body of literature that theorizes contemporary depoliticization in terms of post-politics. The concept of post-politics has emerged from current debates in critical geography and urban studies. According to Swyngedouw whose research focused on identifying the symptoms of the post-political condition within environmental politics, post-politics refers to "the contested and uneven process by which consensual governance of contentious public affairs through the mobilization of techno-managerial dispositives sutures or colonizes the space of the political" (Swyngedouw 2019, xv). Although there is divergence over how to define it (Wilson and Swyngedouw 2015, 7), proponents of the post-political critique such as Jacques Rancière or Slavoj Žižek agree that the concept helps to assess the ways in which our ability to perceive opportunities for change is undermined by those in power.

The post-political framework emerged in the liberal democratic contexts of Europe and America but is increasingly recognized for its potential to unravel state-society interactions in processes of urban governance in Asian cities (Lam-Knott et al. 2019). In this book, the focus is on the processes implemented by the CCP to diffuse agonistic opposition through sustainability practices in the urban sphere. My argument being that Chinese cities increasingly fall into the post-political discussions rendered by foreign scholars. Similarly to what geographers such as Maria Kaika, David Harvey, or Erik Swyngedouw depict in European settings, in China, the layout of the city is groomed by the state to reproduce neoliberal forms of production and hegemony aimed at improving competitiveness and power. As I develop in the following chapters, these mechanisms are increasingly common in Chinese urban spaces such as Shanghai. Yet, I use the post-political concept not to question "post-democratic institutional arrangements" in liberal-democratic settings, but to interrogate the CCP's new modes of exercising power through consensual narratives that value "sustainability" in urban spaces, asking: How do common practices of life in urban spaces come to bear the imprint of an *environmental authoritarian* model of governance?

As stressed above, sustainability has become an integral part of China's guiding political ideology. But current literature offers few detailed considerations of the heterogeneous relations through which policies of sustainable urban development are being mobilized and assembled to regulate everyday citizen behaviors. To assess the CCP's growing instrumentalization of sustainability, and the role of SGOs in materializing those narratives—how they create post-political urban environments—the following book uses *actor-network theory*, commonly abbreviated to ANT, and known as the sociology of translation. An *actor-network* perspective is used to investigate Xi's *environmental authoritarian* approach to governance and how it enhances the creation of a "green" citizenship. More specifically, I will explore how this approach helps the government to resolve governance issues, dissipate confrontational attitudes towards the state and enhance China's position in the race for global recognition.

The ANT methodological approach holds that everything in the social and natural world exists in constantly shifting networks and is thus useful for questioning whether China's "green" consensus is being framed in such a way that it cannot be contested. From the diverse range of SGOs I encountered in the field, to which I will refer when necessary, I choose to take two as case studies: (1) ZeroWaste, an environmental NGO focused on waste sorting; and (2) Farming, a social enterprise that supports local farmers and sustainable farming. These two case studies provide polar cases, and so will be valuable to assess how post-political environments are being spread across various spatial contexts through different relations between different actors. "Following the actors" and assessing how the relations between them are constructed helps us conceptualize the practices of *translation*—how entities' interests, goals, or desires are represented, simplified, and transformed into facts—and, therefore, redirect the analysis to the processes by which *environmental authoritarianism* is constructed, negotiated, and maintained.

Yet although ANT's "flat ontology" is useful to describe the various actors at play in an authoritarian regime such as China, its horizontal perspective doesn't reveal power asymmetries. To recalibrate the significance of power, in this book I therefore combine ANT with Foucault's concept of *governmentality*. *Governmentality*—the governing of people's conduct through positive means rather than only hierarchical top-down enforcement—is used to illustrate how the state objectives require "soft" strategies and how cooperative projects with ZeroWaste or the continuous push for marketisation reinforce the party-state's visions and ideas at the grassroots level.

It is also worth asserting that the research presented here posits itself against the "dichotomous stereotyping" of China's civil society (Salmenkari 2017) that occurs if one neglects to recognize that an antagonistic relation exists between state and society. As such, I do not analyze SGOs through a "political opposition" lens, but as collective activities that aim to achieve certain goals, whereby some ending up assisting the state, while others challenge it.

A note on methodology

My aim, again, is to reflect the on-the-ground reality of urban environmental governance. Building on the work of post-political scholars, I rethink the "political" as involving a multitude of actors with whom power is continually (re)negotiated not only within formal institutions but within the realm of everyday life. Questions of methodology thus arise, including: what method could be appropriate to assess the (re)negotiations of power relations in Shanghai? How to cope with *in situ* constraints? How to account for the socio-political and historical specificities of Shanghai?

Given that I conducted the extended fieldwork during a period of rapid transformation, I have opted for an inductive approach. This entailed, in my case, encountering unpredicted results. I departed into the field in 2016 with an optimistic (and maybe naïve) view of China's civil society development perspectives, explained by the exponential growth of SGOs and misleading interpretations of their potential in a path towards democracy or, at least, the weakening of the state (Balme 2013; Florence 2014).

It is worth noting that, in an initial phase, I aimed to focus on "grassroots" Chinese environmental SGOs. Quickly, though, I realized that this would only reveal a tiny part of Shanghai's organizational diversity. Besides adding constraints, the increasing institutionalization of China's third sector was changing both Chinese and foreign SGOs' action strategies. Between the time I planned my fieldwork and the actual fieldwork phase, the adaptation of SGOs to these new constraints became the hot topic on the ground. Consequently, and given the great importance that foreign organizations have in Shanghai, I broadened my analysis to include foreign environmental SGOs. I mainly developed this phase during a first exploratory phase of three months in Shanghai, from April to June 2016.

During this first phase of analysis, SGOs were uncertain about how hard the Charity Law and the Overseas NGO Management Law would hit them (there is more detail on this in the following chapter). Snowball sampling and my affiliation to a Chinese University revealed to be crucial tools to overcome fear and mistrust. It was through contacts of my Chinese colleagues that I was able to gain access to the Worldwide Fund for Nature representatives. From this first contact, I could then reach out to ZeroWaste and other Chinese SGOs representatives who were not otherwise keen to meet a foreign researcher in such an uncertain period. Nonetheless, because of political sensitivity, I have anonymized the names of all the organizations and any informants mentioned here. During the first phase of research, I conducted six semi-structured interviews with founders and representatives of what would later become our case studies (for more details on my sources, selection criteria, and method, please refer to Appendix A).

Then, I conducted a second round of semi-structured interviews from September 2016 to June 2017. In this second phase, documentary and sociological ethnography supplemented my qualitative data analysis. This comprised oral histories, informal conversations, and observations. I collected and analyzed reports, books, artefacts, photographs, propaganda posters and, more importantly, involved myself in the daily activities of the SGOs under analysis. Slowly, an extended time in the field enabled me to create bonds of trust with my informants who included SGO leaders, workers, scholars, or volunteers, among others. Immersing myself in the field was crucial to mitigate the effects of my presence and reveal the underpinnings behind the official discourse of SGOs, revealing the difference between information shared during formal interviews, and the reality on the ground.

By volunteering and actively taking part in ZeroWaste or BlueOcean activities I could observe the organizations' goals, how they achieved them and, more importantly, see the local effects and insights they received in the communities. As my ability to access the field became smoother, my informants became less constrained when speaking. Deep and long periods of observation, detailed field notes, and perhaps most importantly, *in situ* informal conversations were extremely helpful in allowing me to conduct more direct and focused, oriented interviews on the one hand, and gain insights into the dynamics, processes and strategies of the organizations under study on the other. These spontaneous exchanges opened spaces to touch upon sensible topics, such as government pressures and influences, difficult relations with *shequ* 社区 (community) or property management delegates, but also intra-organizational dynamics and tensions.

I complemented the data gathered in the "real-world" environment with digital ethnography. In addition, I analyzed and took part in daily WeChat forums conducted by the organizations' and volunteers' private groups. As social networks become part of everyday life, experiencing how my informants behave in context and in-the-moment, helped me to bypass the constraints inherent to the institutionalization of SGOs in an authoritarian regime while mitigating the effects of being perceived as a "foreigner".

The chapters ahead

In confronting the implications of the instrumentalization of the environmental sphere by China's authoritarian regime, it is essential to contemplate the different layers that presuppose such an interest for the Party-state. Following this introduction, Chapter 2, "'Greening' authoritarianism", introduces the hallmarks of *environmental authoritarianism*. It particularly explores how the CCP has been developing a model of environmental governance based on both coercive and collaborative mechanisms. Then, the chapter moves on to further develop a conceptual framework by combining a literature on *environmental authoritarian* governance with ANT.

Chapter 3, "The cooperative road towards sustainability in Shanghai", sets the scene and critiques the apparent depoliticization identified in Chapter 2 in the context of Shanghai, identifying the assemblages of diverse actors taking on new roles in response to the many environmental and societal challenges the city needs to address. Centering on the metropolis's unique historical, social, economic, and social settings, I direct attention towards the reasons leading Shanghai to incorporate new social actors into the arena of governing.

Chapter 4, "An iron fist in a velvet glove'" reflects on the role of SGOs in the proliferation of the CCP's substantive speech on the need to protect the environment and call for more public participation. It particularly shows how a state-SGO cooperative approach to waste management is being used to fill the vacuum in public service provision created by years of continuous decentralization and neoliberal economic reform, thereby creating a "green" citizenship aimed at diminishing the political space of resistance and emancipation. This chapter shows empirical evidence of how the CCP engages in cooperative governance strategies, with the help of SGOs, to shape the "conduct of conduct" ("conduire des conduites") of citizens (Foucault 2001).

Chapter 5, "Embracing the market", focuses on the personal experience of Song (not his real name), a former English teacher and real-estate consultant who went back to live in his hometown in Hunan. The chapter explores how, to counter restrictions or avoid institutionalization, his organization, Farming, redefines its boundaries by embracing market strategies and engaging with diverse partners. Specifically, the chapter stresses that, even in a regime as authoritarian as China, global narratives do influence SGOs' development prospects. Yet, although it is tempting to consider the marketisation of SGOs as proof of emancipation, it argues that organizations risk falling into ambivalent and apolitical narratives that reaffirm the goals of the party, thus nurturing a model of "sustainable urbanism".

The prospects and inconsistencies of the CCP's authoritarian approach to governance are further explored in Chapter 6, "Urban sustainability as consensual practice". The chapter explores the underbelly of the CCP's growing green discourses by questioning how cities are produced. It stresses that Shanghai's "sustainable" planning is cultivating active and responsible citizens while rehabilitating a bounded community framework whereas the party-state keeps its central role in community building. Ultimately, this chapter conceptualizes the rise of an "ecological civilization" governance model as a particular form of the post-political.

The final chapter wraps up the book's principal arguments and analyses.

Notes

1 When asked about what is more important, economic development or environmental protection, 11.2 percent of the 3,400 respondents in that study chose the first option.
2 Nineteenth National Congress of the Communist Party of China, 18 October 2017.
3 On 1 June 2017, the China Internet Security Law was enacted. This requires network operators to store select data within China and allows Chinese authorities to conduct spot-checks on a company's network operations (Wagner 2017).
4 Chinese police broke into the home of a retired Shandong University professor who is critical of China's human rights record during a live telephone interview. The live arrest can be viewed on YouTube: https://www.youtube.com/watch?v=6GkjYmBgdOU (accessed 2 September 2020).
5 For instance, Fulong Wu's work on how market-oriented reform reshapes Chinese citizenship.

References

Allmendinger, Phil, and Graham Haughton. 2012. "Post-Political Spatial Planning in England: A Crisis of Consensus?" *Transactions of the Institute of British Geographers* 37 (1): 89–103.

Arantes, Virginie. 2016. *Analyse d'un Mouvement Anti-Barrage En Chine: Conflits, Controverses et Réseau d'actants.* Brussels Working Papers.

Balme, Stephanie. 2013. *La Tentation de La Chine– Nouvelles Idées Reçues Sur Un Pays En Mutation.* Paris: Le Cavalier Bleu.

Beeson, Mark. 2010. "The Coming of Environmental Authoritarianism". *Environmental Politics* 19 (2): 276–294.

Beeson, Mark. 2016. "Environmental Authoritarianism and China". In *The Oxford Handbook of Environmental Political Theory,* edited by Teena Gabrielson, Cheryl Hall, John M. Meyer, and David Schlosberg, 520–532. Oxford University Press.

Beeson, Mark. 2018. "Coming to Terms with the Authoritarian Alternative: The Implications and Motivations of China's Environmental Policies". *Asia and the Pacific Policy Studies* 5 (1): 34–46.

Beeson, Mark. 2019. *Rethinking Global Governance.* Red Globe Press.

Boyd, Olivia. 2013. "What Happened to China's Environment in 2013?" *China Dialogue.* https://chinadialogue.net/en/pollution/6586-what-happened-to-china-s-environment-in-2-13.

Brain, Stephen, and Viktor Pál. 2018. *Environmentalism under Authoritarian Regimes.* Routledge.

Brenner, Neil, and Nik Theodore. 2002. *Spaces of Neoliberalism: Urban Restructuring in North America and Western Europe.* Blackwell.

Brettell, Anna M. 2003. "The Politics of Public Participation and the Emergence of Environmental Proto-Movements in China". University of Maryland, PhD Dissertation.

Brunner, Elizabeth. 2019a. *Environmental Activism, Social Media, and Protest in China: Becoming Activists Over Wild Public Networks.* Lexington Books.

Brunner, Elizabeth. 2019b. "Masked Demonstrations: Deploying Creative Tactics to Protest Air Pollution". In *China's Youth Cultures and Collective Spaces: Creativity, Sociality, Identity and Resistance*, edited by Vanessa Frangville and Gaffric Gwennaël, 135–149. Routledge.

Bruun, Ole. 2017. "Climate, Environment and State–Society Relations in the Mobilisation for Welfare in China". In *Handbook of Welfare in China*, edited by Beatriz Carrilo, Johanna Hood, and Paul I Kadetz, 363–388. Edward Elgar Publishing.

Buxton, Nick. 2017. *Understanding and Confronting Authoritarianism.* Workshop report, War and Pacification Project, Amsterdam.

Carmin, JoAnn, and Adam Fagan. 2010. "Environmental Mobilisation and Organisations in Post-Socialist Europe and the Former Soviet Union". *Environmental Politics* 19 (5): 689–707.

Chen, Geoffrey C., and Charles Lees. 2018. "The New, Green, Urbanization in China: Between Authoritarian Environmentalism and Decentralization". *Chinese Political Science Review* 3 (2): 212–231.

Childe, V. Gordon. 1959. "Urban Revolution". *The Town Planning Review* 21 (1): 3–17.

de Certeau, Michel. 1984. "Walking in the City". In *The Practice of Everyday Life*, 91–110. University of California Press.

de Jong, Martin, Dong Wang, and Chang Yu. 2013. "Exploring the Relevance of the Eco-City Concept in China: The Case of Shenzhen Sino-Dutch Low Carbon City". *Journal of Urban Technology* 20 (1): 95–113.

Della Porta, Donatella, and Mario Diani. 2020. *Social Movements: An Introduction. Contemporary Sociology.* Vol. 29. Wiley-Blackwell.

Doherty, Brian, and Marius de Geus. 1996. *Democracy and Green Political Thought: Sustainability, Rights, and Citizenship.* Routledge.

Doyle, Timothy, and Adam Simpson. 2007. "Traversing More than Speed Bumps: Green Politics under Authoritarian Regimes in Burma and Iran". *Environmental Politics* 15 (5): 750–767.

Dryzek, John. 1987. *Rational Ecology: Environment and Political Economy.* Basil Blackwell.

Eaton, Sarah, and Genia Kostka. 2014. "Authoritarian Environmentalism Undermined? Local Leaders' Time Horizons and Environmental Policy Implementation in China". *The China Quarterly* 218 (1): 359–380.

Eckersley, Robyn. 1992. *Environmentalism and Political Theory: Toward an Ecocentric Approach.* SUNY Press.

Economy, Elizabeth. 2010. *The River Runs Black: The Environmental Challenge to China's Future.* Cornell University Press, 2nd edition.

Elfstrom, Manfred. 2021. *Workers and Change in China: Resistance, Repression, Responsiveness.* Cambridge University Press.

Florence, Éric. 2014. "China, 1978–2013: From One Plenum to Another. Reflections on Hopes and Constraints for Reform in the Xi Jingping Era". *Madariaga* 7 (7) (July).

Foucault, Michel. 1975. *Surveiller et Punir*. Gallimard.

Foucault, Michel. 2001. *Dits et Écrits, I 1954–1975*. Gallimard.

Gauss, Allison. 2016. "Is It Time to Ditch the Word 'Nonprofit'?" *Standford Social Innovation Review*. https://ssir.org/articles/entry/is_it_time_to_ditch_the_word_nonprofit#.

Geall, Sam. 2013. *China and the Environment: The Green Revolution*. Zed Books Ltd.

Giugni, Marco, and Maria T. Grasso. 2015. "Environmental Movements in Advanced Industrial Democracies: Heterogeneity, Transformation, and Institutionalization". *Annual Review of Environment and Resources* 40: 337–361.

Hamnett, Chris. 2020. "Is Chinese Urbanisation Unique?" *Urban Studies* 57 (3): 690–700.

Han, Heejin. 2015. "Authoritarian Environmentalism under Democracy: Korea's River Restoration Project". *Environmental Politics* 24 (5): 810–829.

Han, Heejin. 2017. "Singapore, a Garden City: Authoritarian Environmentalism in a Developmental State". *Journal of Environment and Development* 26 (1): 3–24.

Hansen, Mette Halskov, and Zhaohui Liu. 2018. "Air Pollution and Grassroots Echoes of 'Ecological Civilization' in Rural China". *The China Quarterly* 234: 320–339.

He, Shenjing, and C. S. George Lin. 2015. "Producing and Consuming China's New Urban Space: State, Market and Society". *Urban Studies* 52 (15): 2757–2773.

Hicks, Barbara E. 1996. *Environmental Politics in Poland: A Social Movement between Regime and Opposition*. Columbia University Press.

Ho, Peter. 2007. "Embedded Activism and Political Change in a Semiauthoritarian Context". *China Information* 21 (2): 187–209.

Hsu, Carolyn. 2010. "Beyond Civil Society: An Organizational Perspective on State–NGO Relations in the People's Republic of China". *Journal of Civil Society* 6 (3): 259–277.

Hsu, Jennifer Y.J. 2014. "Chinese non-governmental organisations and civil society: A review of the literature". *Geography Compass* 8 (2): 98–110.

Hsu, Jennifer Y.J. J, and Reza Hasmath. 2015. *NGO Governance and Management in China*. Routledge.

Johnson, Thomas, Anna Lora-Wainwright, and Jixia Lu. 2018. "The Quest for Environmental Justice in China: Citizen Participation and the Rural–Urban Network against Panguanying's Waste Incinerator". *Sustainability Science* 13 (3): 733–746.

Kang, Shih-Hao, and Dayong Hong. 2010. "Zhongguo Minjian Huanbao Liliang de Chengzhang 中國民間環保力量的成長" [The Growing Nongovernmental Forces for Environmental Protection in China]. *East Asian Science, Technology and Society: An International Journal* 4 (3): 457–460.

Khrushchev, Nikita S., and Mikhail Gorbachev. 1993. "The Environmental Movement and Environmental Politics". In *Troubled Lands: The Legacy Of Soviet Environmental Destruction*, edited by D. J. Peterson, 193–228. Westview Press Inc.

Kostka, Genia, and Chunman Zhang. 2018. "Tightening the Grip: Environmental Governance under Xi Jinping". *Environmental Politics* 27 (5): 769–781.

Krueger, Rob, and David Gibbs. 2007. *The Sustainable Development Paradox: Urban Political Economy in the United States and Europe*. Guilford Press.

Lam-Knott, Sonia, Creighton Connolly, and Kong Chong Ho. 2019. *Post-Politics and Civil Society in Asian Cities: Spaces of Depoliticisation*. Routledge.

Latour, Bruno. 2017. *Où Atterrir? Comment s'orienter En Politique*. La Découverte.

Lee, Kingsyhon, and Ming-Sho Ho. 2014. "The Maoming Anti-PX Protest of 2014: An Environmental Movement in Contemporary China". *China Perspectives* 3: 33–39.

Lefebvre, Henri. 1970. *La Révolution Urbaine*. Gallimard.

Lefebvre, Henri. 2000. *La Production de l'Espace*. Economica.

Lei, Ya-Wen. 2017. *The Contentious Public Sphere: Law, Media, and Authoritarian Rule in China*. Princeton University Press.

Li, Huifeng, and Martin de Jong. 2017. "Citizen Participation in China's Eco-City Development. Will 'New-Type Urbanization' Generate a Breakthrough in Realizing It?" *Journal of Cleaner Production* 162: 1085–1094.

Li, Yifei, and Judith Shapiro. 2020. *China Goes Green: Coercive Environmentalism for a Troubled Planet*. Polity.

Lin, George C. S. 2007. "Chinese Urbanism in Question: State, Society, and the Reproduction of Urban Spaces". *Urban Geography* 28 (1): 7–29.

Lin, Jing, and Heidi Ross. 2005. "Addressing Urgent Needs: The Emergence of Environmental Education in China". *China Environment Series* 7: 74–78.

Lin, Teh Chang. 2007. "Environmental NGOs and the Anti-Dam Movements in China: A Social Movement with Chinese Characteristics". *Issues and Studies* 43 (4): 149–184.

Lora-Wainwright, Anna. 2017. *Resigned Activism: Living with Pollution in Rural China*. MIT Press.

Lovell, Hogan. 2020. *Maoism: A Global History*. Vintage.

Luger, Jason. 2016. "Comparative Urbanism and the Authoritarian City: Complexity, Context, and Strategic-Relationality". https://www.researchgate.net/publication/305720958_Comparative_Urbanism_and_the_Authoritarian_City_Complexity_Context_and_Strategic-Relationality?channel=doi&linkId=579baa1708ae5d5e1e1381a1&showFulltext=true.

Luger, Jason. 2018. "Enter the Planetary Authoritarian City? Implications for Research and Theory". Working paper. https://www.researchgate.net/publication/327011388_Enter_the_Planetary_Authoritarian_City.

Ma, Qiusha. 2004. *Classification, Regulation, and Managerial Structure: a Preliminary Inquiry into the Governance of Chinese NGOs*. In *Social Capital in Asian Sustainable Development Managament: Examples and Lessons*, edited by Samiul Hasan and Mark Lyons. Nova Science Publishers.

Moore, Scott M. 2014. "Modernisation, Authoritarianism, and the Environment: The Politics of China's South–North Water Transfer Project". *Environmental Politics* 23 (6): 947–964.

Najam, Adil. 1996. "NGO Accountability: A Conceptual Framework". *Development Policy Review* 14 (4): 339–354.

O'Brien, Kevin J., and Yanhua Deng. 2017. "Preventing Protest One Person at a Time: Psychological Coercion and Relational Repression in China". *China Review* 17 (2): 179–201.

Oakes, Tim. 2019. "China's Urban Ideology: New Towns, Creation Cities, and Contested Landscapes of Memory". *Eurasian Geography and Economics* 60 (4): 400–421.

Ophuls, William. 1973. *"Leviathan or Oblivion" towards a steady state economy*. W. H. Freeman.

Ophuls, William, and Stephen A. Boyan. 1992. *Ecology and the Politics of Scarcity Revisited: The Unravelling of the American Dream. Trends in Ecology & Evolution*. W. H. Freeman.

Pei, Minxin. 2013. "Five Ways China Could Become a Democracy". *The Diplomat*. https://thediplomat.com/2013/02/5-ways-china-could-become-a-democracy.

Perry, Suzanne. 2009. "Goodbye, 'Nonprofit Sector'. Hello, 'Delta Sector'". *The Chronice of Philanthropy*. https://www.philanthropy.com/article/goodbye-nonprofit-sector-hello-delta-sector/.

Pinson, Gilles, and Christelle Morel Journel. 2016. "The Neoliberal City—Theory, Evidence, Debates". *Territory, Politics, Governance* 4 (2): 137–153.

Pow, Choon-Piew. 2009. *Gated Communities in China: Class, Privilege and the Moral Politics of the Good Life*. Routledge.

Reese, Stephen D. 2015. "Globalization of Mediated Spaces: The Case of Transnational Environmentalism in China". *International Journal of Communication* 9: 19.

Saich, Tony. 2000. "Negotiating the State: The Development of Social Organizations in China". *The China Quarterly* 161 (February): 124–141.

Salamon, Lester M., Helmut K. Anheier, Regina List, Stefan Toepler, and Wojciech Sokolowski. 1999. *Global Civil Society: Dimensions of the Nonprofit Sector* (Vol. 2). Kumarian Press, Inc.

Salmenkari, Taru. 2017. *Civil Society in China and Taiwan*. Routledge.

Sarin, Aaron. 2020. "The Crimes of the Red Emperor". Quillette. https://quillette.com/2020/08/31/the-crimes-of-the-red-emperor/#_ednref1.

Schindler, Seth. 2017. "Towards a Paradigm of Southern Urbanism". *City* 21 (1): 47–64.

Shabar, Dan Coby. 2015. "Rejecting Eco-Authoritarianism, Again". *Environmental Values* 24 (3): 345–366.

Shapiro, Judith. 2013. "The Evolving Tactics of China's Green Movement". *Current History* 112 (755): 224.

Shieh, Shawn; Schwartz, Jonathan. 2009. *State and Society Responses to Social Welfare Needs in China: Serving the People*. Routledge.

Sima, Yangzi. 2011. "Grassroots Environmental Activism and the Internet: Constructing a Green Public Sphere in China". *Asian Studies Review* 35 (4): 477–497.

Simone, Abdoumaliq. 2014. *Jakarta, Drawing the City Near*. University of Minnesota Press.

Simpson, John. 2016. Critics Fear Beijing's Sharp Turn to Authoritarianism. *BBC News*. https://www.bbc.co.uk/news/world-35714031.

Solinger, Dorothy J. 1999. *Contesting Citizenship in Urban China: Peasant Migrants, the State, and the Logic of the Market*. University of California Press.

Sowers, Jeannie. 2007. "Nature Reserves and Authoritarian Rule in Egypt: Embedded Autonomy Revisited". *Journal of Environment and Development* 16 (4): 375–397.

Spires, Anthony J., Lin Tao, and Kin-man Chan. 2014. "Societal Support for China's Grass-Roots NGOs: Evidence from Yunnan, Guangdong and Beijing". *The China Journal* 1 (71): 65–90.

Stalley, Phillip, and Dongning Yang. 2006. "An Emerging Environmental Movement in China?" *The China Quarterly* 186: 333–356.

Steinhardt, H. Christoph. 2019. "Environmental Public Interest Campaigns: A New Phenomenon in China's Contentious Politics". In *Handbook of Protest and Resistance in China*, edited by Teresa Wright, 235–252. Handbooks, 480. Edward Elgar Publishing.

Swyngedouw, Erik. 2011. "Whose Environment? The End of Nature, Climate Change and the Process of Post-Politicization". *Ambiente & Sociedade* 14 (2): 69–87.
Swyngedouw, Erik. 2019. *Promises of the Political: Insurgent Cities in a Post-Political Environment*. MIT Press.
Tai, John W. 2015. "Chinese NGOs: Thriving Amidst Adversity". In *Building Civil Society in Authoritarian China*, 20: 19–43. Springer.
Tuan, Yi-Fu. 1989. "Surface Phenomena and Aesthetic Experience". *Annals of the Association of American Geographers* 79 (2): 233–241.
van Rooij, Benjamin, Rachel E. Stern, and Kathinka Fürst. 2016. "The Authoritarian Logic of Regulatory Pluralism: Understanding China's New Environmental Actors". *Regulation and Governance* 10 (1): 3–13.
Wagner, Jack. 2017. China's Cybersecurity Law: What You Need to Know. *The Diplomat*. https://thediplomat.com/2017/06/chinas-cybersecurity-law-what-you-need-to-know.
Wang, Shaoguang, and Jianyu He. 2004. "Associational Revolution in China: Mapping the Landscapes". *Korea Observer* 35 (3): 485–534.
Wells, Peter. 2007. "The Green Junta: Or, Is Democracy Sustainable?" *International Journal of Environment and Sustainable Development* 6 (2): 208–220.
Wilson, Japhy, and Erik Swyngedouw. 2015. *The Post-Political and Its Discontents: Spaces of Depoliticisation, Spectres of Radical Politics*. Edinburgh University Press.
Wong, Kit Ping. 2019. "Territorially-Nested Urbanization in China—the Case of Dongguan". *Eurasian Geography and Economics* 60 (4): 486–509.
Wong, Natalie W.M. 2016. "Environmental Protests and NIMBY Activism: Local Politics and Waste Management in Beijing and Guangzhou". *China Information* 30 (2): 143–164.
Wu, Fengshi, and Bo Wen. 2014. "Nongovernmental Organizations and Environmental Protests: Impacts in East Asia". In *Routledge Handbook of Environment and Society in Asia*, edited by Paul G. Harris and Graeme Lang, 121–135. Routledge.
Wu, Fulong. 2010. "Property Rights, Citizenship and the Making of the New Poor in Urban China". In *Marginalization in Urban China*, edited by Fulong Wu and Chris Webster, 72–89. Springer.
Wu, Fulong. 2016. "State Dominance in Urban Redevelopment: Beyond Gentrification in Urban China". *Urban Affairs Review* 52 (5): 631–658.
Wu, Fulong. 2020. "Adding New Narratives to the Urban Imagination: An Introduction to 'New Directions of Urban Studies in China'". *Urban Studies* 57 (3): 459–472.
Xie, Lei. 2009. *Environmental Activism in China*. Routledge.
Xie, Lei. 2011. "China's Environmental Activism in the Age of Globalization". *Asian Politics & Policy* 3 (2): 207–224.
Xie, Lei. 2015. "Political Participation and Environmental Movements in China". In *The International Handbook of Political Ecology*, edited by Raymond L. Bryant, 246–259. Edward Elgar Publishing.
Xie, Lei, and Peter Ho. 2008. "Urban Environmentalism and Activists' Networks in China: The Cases of Xiangfan and Shanghai". *Conservation and Society* 6 (2): 141.
Ye, Lin. 2017. *Urbanization and Urban Governance in China: Issues, Challenges, and Development*. Springer.
Zhong, Yang, and Wonjae Hwang. 2015. "Pollution, Institutions and Street Protests in Urban China". *Journal of Contemporary China* 25 (98): 216–232.

2 “Greening” authoritarianism

Introduction

With millions of subscribers on YouTube and Weibo, Li Ziqi 李子柒, a Chinese internet celebrity, has become a global sensation. Her fairy-tale-like depictions of the delicate art of Shu embroidery or mundane farm work set against a scenic mountain backdrop ignited a strong interest from netizens in China and abroad. For many of her followers, Li Ziqi’s videos represent an escape from hectic, fast-paced urban living. Even though the vlogger’s intentions remain a mystery, sparking vigorous debate on the net, her beautiful and polished images of China’s traditional rural lifestyle have reached millions of people around the globe. Widely active on YouTube, despite the platform being officially blocked in China, many observers advocate for her to receive government support.

Given China’s state-monitored public discourse, any chance to showcase an image of China that doesn’t align with that of the party on a western media would be short-lived without the consent of the state. One need only recall how Fang Fang’s online Wuhan diary, depicting the life of the people during the COVID-19 pandemic, was censured after its author was accused of spreading a “doomsday narrative”. For Jessica Imbach (2020), Li’s videos, in contrast to Fang Fang’s writing, promote China’s conceptual framework of “ecological civilization” because they connect environmental imaginaries within a Chinese cultural body. On the one hand, the videos serve the CCP’s efforts to promote an alternative sustainable path to the Western lifestyles and, on the other, they normalize the omnipresence of nationalist sentiments (Wang 2021). This contrast is the new reality in China: the authorities adopt a firm hand with voices that disagree or show China’s path of development in a negative light, while promoting bottom-up representations that feature China’s traditional cultural values of harmony, honesty, filial piety, and ecological wisdom.

This dichotomy is what this chapter is about. When Xi Jinping took the reins of power in 2013, the CCP stepped up its efforts to eradicate negative images of China. The Publicity Department of the Central Committee of the Communist Party of China, also known as the Propaganda Department, for

DOI: 10.4324/9781003231325-2

instance, has increasingly shut down internet celebrities and imposed tighter regulations. This "cyber sovereignty" (Tager et al. 2018) has also affected the emergence of environmental movements by erasing and controlling undesirable or sensitive information from the public domain. The government's resumption of the Tiger Leaping Gorge dam construction is a good example of the direct consequences of these measures. Although construction had previously been blocked by a robust environmental movement (as described in the first lines of the book), the absence of opposition now symbolizes how much has changed since China's parliament elected Xi Jinping president.

In this chapter, I aim to address these challenges and questions, as well as gauge the changing strategies used by the Chinese state to deal with the environmental question. The question is, how has the state machinery slowly co-opted environmental issues, long interpreted in the literature as favorable for the development of Chinese civil society (Xie and Ho 2009; Xie and Van Der Heijden 2010)? The chapter starts with a historical perspective by succinctly outlining how the CCP has responded to the rise of environmental movements since the beginning of the 2000s. Next, I present the theoretical underpinnings of the book, specifically why I choose to combine the conceptual tools of *environmental authoritarianism* with Actor-Network Theory (ANT). Here I will argue that governance mechanisms more typical of democratic regimes are being used by Chinese leaders to stabilize and strengthen authoritarian rule in urban settings. This chapter thus also introduces one of the major arguments of the book, namely the ability of the party-state to co-opt environmental politics with a view to pursuing other strategic goals.

2000s–2010s: Looking for "blue skies"

The Baiji dolphin is the largest mammal living in the Yangtze, China's longest river. Or at least, it used to be. Estimated to have been present in the Yangtze River for 20 million years, scientists refer to Baiji as the first dolphin species driven to extinction due to human activity. Its demise was rapid and shocking, going from a healthy population of some 6,000 animals to become extinct in just a few decades.[1] This mythic species, which seemed to disappear in the blink of an eye, symbolizes the devastating environmental price that China has paid after decades of uncontrolled development. In 2003, the government issued an annual fishing ban lasting three months every year. They extended this to four months per year from 2016 to protect fish stocks, and in 2021, they outright banned fishing in the country's crucial river.[2]

It is not news that China's economic boom, despite its positive effects on living standards, has transformed the country's environment. The massive industrialization process damaged biodiversity and plunged the entire country into a tangible ecological crisis. China's Gross Domestic Product (GDP), which grew 10 percent per year on average for more than a decade, was achieved at the expense of the environment and public health. The country's economic output has left a severe toxic legacy: air, soil and water pollution, food, and energy security. At the start of the 2010s, around 1.6 million people died from air pollution every year (Rohde and Muller 2015). The threat affected city-dwellers' health in the country's biggest cities (Kan 2009), resulting in a decrease of 5.5 years of life expectancy in the country's north compared to other less developed areas in the south (Y. Chen et al. 2013).

China's mounting environmental crisis thus became one of the most pressing challenges to the legitimacy of the Party-State, both domestically and abroad. The measures taken by the regime to avoid being remembered for the "Smoglympics" in 2008 encapsulate how central the issue had become for the party. To reach a "Blue sky", the Olympic village retrofitted or closed nearly 200 factories, restricted car usage and slowed down construction projects. It had become so crucial for the party to showcase Beijing's "blue skies" that the city's weather engineers shot silver iodide into incoming clouds to ensure the rain would be flushed out before reaching the Bird's Nest. Despite these efforts, however, dangerous levels of smog persisted. Conditions were so severe that some athletes trained in neighboring countries and travelled to Beijing before the opening ceremony.[3] Worst, suspicions prevailed as the levels of pollution readings released by the authorities showed inconsistencies when compared to measurements provided by the Associated Press.[4]

Chinese leaders were here given the opportunity to showcase the country's economic evolution and political power, but the world was left with the image of grey skies, smog, face masks, and lack of transparency. The alarming situation in China even resulted in terms never heard before, such as "cancer villages", as reported by some of the largest broadcasters in the world such as the BBC and the *Guardian*.[5]

While public anger rose in the north of the country over air pollution, there were also extensive discussion surrounding water supplies, accused of causing the rate of cancers to rise at alarming rates. Even though Chinese authorities spent millions on the issue,[6] 85 percent of the water in the city's major rivers was undrinkable in 2015, according to official standards, and 56.4 percent was unfit for any purpose.[7] According to the same sources, around 70 percent of the country's rivers and lakes were severely polluted. Moreover, water pollution is not an isolated problem. It also damages soil, with untreated wastewater contaminating soil and pollutants in surface and groundwater sources. It also threatens food security, and this vicious cycle saw the emergence of food scandals all over the country. A scandal over Chinese milk in 2008 saw six infants died from kidney damage, and became one of the most notorious food safety incidents, while the severe cadmium pollution of rice in China's largest rice-producing province, Hunan, also reached the headlines all over the world.

2010–2015: growing environmental consciousness

Because hazardous levels of water and air pollutants became a daily phenomenon in the lives of the Chinese (especially in developed coastal areas), concerns over environmental issues became one of the population's top concerns in the 2010s, right after corruption of officials, but ahead of wealth inequality, crime, or unemployment (Wike and Parker 2015). Discontent over pollution concerns was particularly high among younger generations, with 60 percent of young respondents in a study by Kong et al. (2014) saying they "completely agreed" that the state should prioritize environmental concerns over economic growth.

The public reactions following the release of *Under the Dome* (*qiongding zhi xia* 穹顶之下), a documentary produced by former CCTV journalist Chai Jing 柴静 in 2015, illustrates how crucial environmental issues became among the Chinese population. The film got such a huge amount of attention that within the space of three weeks, the documentary went from being viral on the internet, then being blocked by government censors, to finally being the subject of a question to Premier Li Keqiang 李克强 at a press conference where he vowed the regime would take tougher action on environmental issues.[8]

Under the Dome unveiled the contradictory roots of environmental protection versus economic development and exposed the role that media attention played in enabling this wake-up call. As Judith Shapiro (2013) explains of this period, as pollution affected people's human health and well-being, groups within Chinese civil society—including environmental groups, environmental journalists, investigative reporters, and activists—became more assertive in voicing their apprehensions. Social media became a useful tool for sharing information and bypassing limitations on public participation and freedom of information.

The film marked a turning point in China's civil society history, and led to a pluralization of voices and associated growth in environmental consciousness among the Chinese population. Some scholars even saw this as evidence of a new chapter beginning for China's civil society (Lu 2007; J. Chen 2010; Kassiola and Guo 2010; Geall 2013; J. Y. J. Hsu 2014), notably the growth of environmental organizations marked by *Friends of Nature*, China's first environmental NGO, established in 1993.

Social good organizations: new players in the rise of an environmental consciousness

> I always have a positive opinion about these organizations (NGOs). The government needs them. I think they're going to be another arm for the government to manage the country.
>
> (Qu Geping, ancient Director of the Environmental Protection Agency, cited in Marcuse, 2011)

The growth of an environmental consciousness among the Chinese population has happened in parallel with the rise of SGOs. I had previously outlined my decision to use the term SGO over Civil Society Organizations (CSO) or Non-Governmental Organization (NGO). I argue it better depicts the observed organizations' diversity while acknowledging that they all share a desire to act for the "common good". Before exploring their characteristics in more depth, I will take a few lines to briefly summarize their development.

1990s–2000s

Since the early 1980s, and particularly during the 1990s and 2000s, regardless of difficult registration requirements and legal ambiguities, there was a transition towards a market-based economy and a gradual retreat of the state in delivering welfare provision, leading to the emergence of a variety of organizations, with varying degrees of attachment to the State, in a wide range of fields (Ma 2005; P. Ho 2007; J. Y. J. Hsu and Hasmath 2017; Snape 2021). "Their growth has been remarkable", stated Professor Liu (fictitious name) in April 2016 when I interviewed her at the very start of my fieldwork. She went on: "When I worked for the environmental education initiative program in Beijing, in 1997, it was the first time for me to know there is something called an NGO … and you know, I worked for a university based in Shanghai".[9]

2000s–2010s

Although they were unknown to most Chinese people at the beginning of the 2000s, over the years SGOs have grown in quantity, scale, and quality (Chao and Onyx 2015). From developed coastal areas to the most inland rural zones, they can now be found everywhere. The word NGO is no longer a secret—I kept hearing it in my discussions with young people engaged in volunteering activities, professors teaching in primary schools and universities, environmental, or social events in the city center or community activities in Shanghai's periphery. Yet since the appearance of the first "NGOs" on the mainland some 40 years ago, new terms have emerged. Terms such as Corporate Social Responsibility (CSR), Non-Profit Organization (NPO), or Social Enterprise (SE) were used interchangeably by actors in the field, sometimes making it difficult to know what people were referring to.

The constraining context surrounding the growth of SGOs explains this ambiguity between definitions. Until recently, organizations had to register under the constraining "dual management system" in which an NGO first needed to get approval from a Professional Supervisory Unit in a similar field before it could register with Ministry of Civil Affairs (Wang 2010; Snape 2021). To overcome these legal barriers, many organizations registered as businesses entities or did not register at all (Fulda 2017). These ambiguous identities made it difficult to estimate their numbers and characteristics. In 2016, Hsu and Hasmath (2017) estimated that there were approximately 440,000 registered NGOs and many more unregistered.[10] The 2018 Report on Social Organizations in China pointed to some 80 million "social organizations". Yet, as Gabriel Corsetti (2019) argues, the term "social organization" refers to a broad range of bodies and is therefore not a trustworthy indicator.

The administrative category of "social organizations", in official statistics, includes social organizations (*shehui tuanti* 社会团体), civil non-enterprise units (*minban fei qiye danwei* 民办非企业单位), and foundations (*jijin hui* 基金会).[11] Yet many people would not consider the majority of these organizations to be NGOs in the commonly used sense of the word. For example, they are not always spontaneously created, self-governing, or independent as described in the term Government Non-Governmental Organizations (GONGOs) (Corsetti 2019). When I use the term SGOs, I thus aim to refer to bottom-up organizations which are formally or informally organized, and self-governing, however with different levels of connection with the state and the market, non-profit, socially oriented, and with the primary goal of promoting social or environmental goals.

2010s–2016

Environmental SGOs have been one of the most conspicuous phenomena in the recent history of civil society in China (Hasan 2015), especially from 2011 to 2015 following the abolishment of the "double management system" (Jie 2006). Although China's authoritarian government arrested and detained activists advocating human rights and Tibetan independence, the authorities tolerated environmental SGOs development, albeit with increasing restrictions.[12] The alarming dimension of environmental issues expanded the political space afforded to green civil society. From 1996 to 2011, the environmental crisis resulted in an increasing number of protests related to environmental issues.[13]

Unlike protests related to sensitive issues such as human or labor rights that were rarely led by official organizations, the country's environmental plight and related effects on economic development fueled the rise of numerous environmental SGOs. The political reform of government, the defects of environmental governance, a growing environmental consciousness, international pressures, organizations' professionalization, the efforts of influential elites, the international dimension of environmental problems (such as the greenhouse effect), and the internet all contributed to the growth of organizations (Hasan 2015).

The state of SGOs after 2016

As outlined in the introduction, in 2016, SGOs approached a critical juncture. In mid-March, the long-awaited Charity Law laid out the legal foundation for the creation of the philanthropy sector. Despite some positive developments (e.g., eliminating taxes on donations), the law also introduced a legal barrier to foreign funding and resources. Shortly after, in January 2017, the Overseas NGO Management Law (*jingwai fei zhengfu zuzhi jingnei huodong guanli fa* 境外非政府组织境内活动管理法) was passed, requiring overseas organizations in China to register, submit to police supervision, and declare the sources of their funding.

These regulations that were tabled early in Xi Jinping's first term as President, took both practitioners and observers by surprise (Shieh 2018b). Setting the tone for his desire to "remake civil society in the Party-state's image" (Shieh 2018a), China's new leader smashed what until then had been interpreted as a favorable period for the growth of civil society in China (Tai 2015).

A lost opportunity?

Various historical examples from around the world have shown that when actors are outside the reach of the government, co-creating knowledge and new ideas that diverge from the state, this puts authoritarian political control under threat. In post-industrial societies, for instance, the presence of non-governmental actors linked environmental concerns to value systems and matters of quality of life (Birnie et al. 2009). It also led environmental movements to act as civil society mobilizers in Asia (M. Ho 2011), Eastern Europe (Hicks 1996), and Latin America (Jacobs 2002). Even though environmental groups developed according to their own modes of democratic transition, most such movements were founded on a close relationship between democratic consolidation and environmental politics. As an illustration, drastic change emerged under South Korea's ruling bloc (in part) as a response to intense "politics of protest" by environmental civil society groups (Kim 2000). Similarly, environmental issues in Brazil have been intrinsically linked to the struggle for citizenship, rights, and government accountability (Jacobs 2002).

The role of environmentalist critiques in regime transition is also found in state-socialist countries of Europe during the second half of the 1980s (Stec 2005). Environmental issues came to stand for everything that was wrong with state socialism, symbolizing problems such as industrialization, unregulated state control over the economy, or lack of transparency, to name but a few (Carmin and Fagan 2010; Corry 2013). As illustrated by the cases of Hungary and Latvia, communist regimes fell apart as environmental movements grew and spread across the socialist states, bringing a new form of "democratic governance" with the implementation of a multi-party parliamentary system, rule of law, and a budding civil society (Steger 2004). Environmental movements enriched liberal democracy and the voice of citizens by exposing environmental injustices, discrimination, and racism (Eckersley 2020, 4). The emancipatory social movements of the 1970s and 1980s forced environmental issues onto national and international political agendas, fortifying the assumption that ecology leads to democracy (Blühdorn 2011, 1–2).

Looking at the growing number of environmental SGOs in China, and the wide literature on democratic transitions (Battěk and Wilson 1985), expectations increased regarding new civic openings (Cooper 2006; Tang and Zhan 2008; J. Chen 2010; Zhan and Tang 2013). Because China's population has grown wealthier in the past 30 years while severe environmental problems have given way to green activism and public demonstrations, several scholars hoped this would lead the Chinese people to yearn for democratic freedoms, rights, and the rule of law (Saich 2000; Yang and Calhoun 2007; Kassiola et al. 2010; Mertha 2011; Hsu 2014).

Yet against the odds, no step toward democracy have occurred. The strength of the Chinese State and the insufficient capacity of SGOs prevented them from achieving large-scale policy changes. Not only did the Party-State learn how to change, modernize, and adapt itself to the new economic and social reality (Cabestan 2014; E. Florence 2014), it also increasingly restricted and regulated the development and action capacity of SGOs, as I will continue to explore in the following chapters.

Was the regime really under threat?

China's "vibrant" environmental activism was never a real threat to the authoritarian regime. Whenever specific crises or danger to participants' health trigger a campaign, anger gets directed to polluting chemical plants, waste treatment facilities, or coal-fired power plants, to name just a few. Yet as soon as activists have reached their goals, movements dissipate, and therefore stay ramified. Even when a protest blocks the implementation of a proposed plan, the project in question will most likely just be moved to a poorer area. Moreover, although the severity of environmental degradation intensifies the population's demand for higher standards of living, citizens rarely point the finger at the role played by the frenetic model of development being followed by the central government. Quite the opposite, in fact. Environmental campaigns may receive state support and, in most cases, environmental movements direct their discontent at local governments or industries, but they rarely question the role of the central government (Tong 2005).

Since the mid-2000s, environmental issues have become the leading source of protest in China (Steinhardt and Wu 2016), either in underdeveloped rural villages (J. Jing 2000) or in developed urban coastal areas (Johnson 2013; Johnson et al. 2018), however, they have never truly threatened the legitimacy of the Party. When movements occur, they are likely to face repression or censorship from authorities. As an illustration, in late 2016, police detained a group of masked Chinese artists protesting over pollution in Chengdu. That same year, police detained protesters in Hubei and Hunan. More recently, in 2019, authorities violently repressed residents of the Yangluo District in Wuhan who had taken to the streets to dispute a waste-to-energy plant.[14] Moreover, the freedom of several actors—judges, environmental NGOs and citizens—to engage in China's environmental governance processes is being gradually confined, and their action reclaimed by governmental instances (van Rooij et al. 2016).

Taking this period of restriction as a site for analysis, this book aims to assess how the CPC prevented large-scale unrest over environmental damages and rights. Yet before exploring this "green" turning point in more depth, it is necessary to give some background on the changes to grassroots participation and repression that occurred during the shift from Hu Jintao's 胡锦涛 (China's President from 2002 to 2012) cautious and gradual political reforms to Xi Jinping's firm grip on power (Fu and Distelhorst 2018).

From "soft" Hu to "strong" Xi

Despite the efforts of the Hu administration to strengthen central authority (Leutert 2018), a "fragmented" (Lieberthal and Oksenberg 1988), "soft" (Cabestan 2004) and "participatory" authoritarianism allowed society to empower itself *vis-à-vis* the state (Kornreich 2016).[15] Hu and Premier Wen Jiabao's 温家宝 series of measures to establish an "harmonious society" failed here to counter the forces of decentralization and rampant marketization (É. Florence and Defraigne 2013). Although concepts such as "people-centered" development (*yirenweiben* 以人为本) invoked the idea of going beyond GDP growth objectives, a gap between the law and its application remained, as officials relied on GDP growth for self-promotion.

Aimed at balancing economic, social, and environmental objectives, President Hu's strategy created lasting tensions between the economy and the environment (Bina 2008), and therefore, "efficient and rapid" economic development remained limited. As an illustration, policy implementation kept partial leverage over Provincial Governments and municipalities, while the decentralization of the Party posed barriers to Hu's "ecological civilization" plan. Xi made use here of institutional (e.g., reinforcing the Party over the state), and informal mechanisms (e.g., laws and regulations reducing local government authority), alongside tightly controlling and regulating social forces (e.g., media or SGOs).

Under Xi, macro control by the central government was strengthened through discipline inspection and the enforcement of various laws and regulations (e.g., NGO Management Law, Anti-terrorist Law, National Security Law) to prevent the spread of corruption and the formation of regional strongholds that had been thriving since China's reform and opening-up policies. Although Xi's style of politics follows a continuity from the past, such as the concept of "socialism with Chinese characteristics" advanced by Deng Xiaoping 邓小平 at the beginning of the 1980s, or, as I stress below, anti-corruption campaigns, one should note the trend towards the centralization of power to strengthen his grip over the nation's top decision-making bodies and respond to the hurdles his predecessor had faced.

Xi promptly enforced discipline as soon as he came to power. Although anti-corruption campaigns had started in the late 1990s under Jiang Zemin 江泽民 (President of the PRC from 1993 to 2003) and at the end of Hu's term in 2009, no leader before Xi had been quite so harsh on re-centralization efforts (Yuen 2014), and Xi's hostility went well beyond his predecessors' efforts to combat both "corrupt flies and tigers" (*laohu cangying yiqu da* 老虎苍蝇一起打), a frequently repeated metaphor targeting both highly placed power leaders, the most prominent "tiger" case involving Bo Xilai 薄熙来, the former politburo member and governor of Chongqing, and petty local cadres.

Reports of such well-publicized anti-corruption campaign led by President Xi coincided with the detention of grassroots activists. In 2013, the police detained Cao Shunli 曹顺利 for trying to take part in the 2013 Universal Periodic Review of China's human rights record at the United Nations Human Rights

Council in Geneva. For several months, authorities denied Cao access to adequate healthcare even though she was seriously ill.[16] Cao died in March 2014. Other appalling examples of the Xi administration's way of dealing with social unrest include the police actions against the Yirenping network in 2014–15, the detention of the "five feminists" in March 2015, and the repression of human rights defense lawyers, just to name a few. As Teng Biao 滕彪, Chinese human rights lawyer, explained: "2013 saw the harshest suppression of civil society in over a decade".[17]

The Chinese government not only targeted SGOs and activists but also enforced stricter restrictions on the internet and the press by keeping "unacceptable" information out of the reach of most Chinese citizens.[18] "Public accounts" on WeChat, China's most popular messaging platform, were closed around this time, and users obliged to register with their "real" names. Meanwhile, it became harder for Chinese researchers to go abroad. As Eva Pils has put it, the CCP is exploring a psychology of fear and performing a "red line illusion" game to keep scholars' alert.[19] Other strategies, such as visa or travel bans, instructions for censorship, threats, monitoring, or surveillance are also being used to restrain researchers' freedom. The regime also requires scholars and intellectuals to go through an ideological evaluation.[20]

Hence, although the CCP tolerated environmental protests for a time, at least compared to individual rights defenders, since 2016, green activists have received increasing attention from the authorities. In 2016, authorities harassed Chongqing activist Xue Renyi 薛仁义, the head of the group Green Leaf Action, for promoting environmental protection, universal and social healthcare.[21] In May 2018, Xue went missing. According to Chinese Human Rights Defenders,[22] on 31 July 2019, authorities reported that the Dianjiang County Detention Centre of Chongqing Municipality had detained Xue for "picking quarrels and provoking trouble". Another active member of Green Leaf Action, Pan Bin 潘斌,[23] was sentenced to four years' imprisonment for the same offence, in December 2018.

Similarly, in 2016, Liu Shu 刘曙, the founder of the Hunan-based environmental NGO Shuguang, was detained for ten days for providing information to an "unidentified contact" (Economy 2018, 179). She had been investigating levels of heavy metal pollution in Dongting Lake[24] and campaigning for victims of environmental pollution. Another activist, Liu Dazhi 刘大志, an advocate for public health and discrimination cases in the public interest advocacy NGO Changsha Funeng, requested government data on tap water quality and biomedical waste. In August 2019, prosecutors of the security bureau in Changsha (Hunan) approved the arrest of Liu Dazhi and two colleagues working at the same organization. Or, coming back to the milk scandal which caused the death of six infants, even though the factory was shut down and relevant officials dismissed, the authorities attacked a group of lawyers who insisted on further punishment (Fei 2016, 96).

What all these cases have in common is that they draw a straight line between environmental issues and advocacy, questioning the logic of the CCP's economic development strategy.[25] Although environmental SGOs have been tolerated for a time, at least compared to other types of organization, more recently every effort has been made to make sure they stay under the control of the CCP. In such circumstances, if activists touch upon sensitive issues—by disclosing information on an imminent danger to health or the natural environment—it can put them in a delicate position (as happened to Dr Li Wenliang 李文亮, whom local authorities punished after he had tried to warn about a disease that looked like SARS-like virus in late December 2019). The few cases presented here prove that challenging the party's legitimacy (transparency, social justice, or accountability) will no longer be tolerated (Mah 2017).

This shows a discontinuity in government reactions to contentious participation from Hu to Xi (Fu and Distelhorst 2018), as campaigns against civil society have become increasingly threatening and are based on proactive repression. As a result, Chinese SGOs and activists are refraining from making radical statements and increasingly professionalizing to focus on non-sensitive issues—recycling, environmental education, or urban greening. Moreover, as I will develop further, these new roles has increasingly put the burden of environmental issues firmly on citizens' shoulders.

Broadly, this instrumentalization and co-optation of environmental issues runs contrary to the prospects of emancipation described in the literature. As we will see in subsequent chapters, Xi Jinping takes advantage of "green" discourses to counter social unrest and popular movements, and eliminate dissensus, place itself, the party, and the state, at the forefront of environmentalism. A big part of this strategy is reflected in his commitment to creating an "ecological civilization", dissected below.

Figure 2.1. Propaganda Poster referring to the "Chinese dream (*zhongguo meng* 中国梦)" showcased in a destruction site in Shanghai.

"Environmental civilization"

First coined by Hu Jintao in 2007, the concept of "environmental civilization" represents Xi's double goal of creating a "beautiful China (*meili zhongguo* 美丽中国)"[26] and achieving the "Chinese dream (*zhongguo meng* 中国梦)".[27] This "green" rhetoric, I argue, is used to justify a more assertive form of governance at the local level. The CCP uses this argument to advance its role in leading the Chinese people towards "socialist ecological progress". It reinforces imaginaries of national development and nation's rejuvenation and thereby taps into a return to the grandeur of past dynasties which reinforces the role of the party.

Scholars have widely assessed the efforts made by the party and the state to show that China is taking the lead in "ecological progress". The Master Plan for Environmental Protection Reforms (*shengtai wenming tizhi gaige zongti fang'an* 生态文明体制改革总体方案), released in September 2015, or the establishment of two new ministries, the Ministry of Ecological Environment and the Ministry of Natural Resources in 2018, for instance, have been thoroughly analyzed as a sign of the CCP's desire to reduce fragmentation in terms of regulatory mandates.[28] The spotlight in the literature has been particularly directed at administrative measures which, according to Kostka and Zhang (2018), show the party-state's effort to reduce inefficiency and conflicts associated with problems with coordinating what has been framed as a fragmented form of authoritarianism (Z. Yang 2013; Brehm and Svensson 2017).

In particular, the politicization of environmental protection by the CCP has caught scholars' attention since "ecological civilization" was elevated as a constitutional principle of the People's Republic. As Hansen and Liu (2018) argue, it reveals China's "green" policy framework of development, showing how top-down imaginaries are (re)framing authoritative standards of control. Focusing on the linguistic choice of the term "civilization", Heidi Wang-Kaeding argues this intertwines with the party's "spiritual civilization" in the wake of the 1989 Tiananmen Protests. According to the author, the concept functions as a strategy for the CCP to persuade the public that it is possible to reach a sustainable future with "Chinese characteristics",[29] freeing the Chinese state from the obligation to follow a Western democratic model.

The way in which Chinese leaders preyed on President Donald Trump's decision to withdraw from the Paris agreement on climate change was also seen as an opportunity for Xi to fill the gap and strengthen China's role in global climate governance. Likened to the term "Anthropocene" in the West, the concept of "ecological civilization" denotes a significant discursive power. It allows a shift from the binary political economy discourses of "growth" against "development". The concept transcends a Chinese version of the Anthropocene (Marinelli 2018, 368–369) because it offers situated understandings and possibilities for alternative ways of exploring eco-social prosperity in China, in cooperation with society (*gongzhong canyu* 公众参与) (State Council of the People's Republic of China 2019).

Still, literature has remained macro-oriented and focused on the state. Hence, environmental governance is not only a matter of the state. It includes market actors, social organizations, and the general public. To untangle the many institutional setups that Chinese leaders are developing to achieve an "ecological civilization", one needs to go beyond policymaking at the meso- and macro-political levels. This is the task I aim to achieve in this book. Rather than focusing on the state level, the subsequent chapters show that China's environmental authoritarian model of governance also unfolds through everyday micro-politics. I will particularly focus on how the model plays out at a micro-community level and colonize city-dwellers' daily lives.

Reviving environmental authoritarianism

Environmental authoritarianism (or eco-authoritarianism) holds that democracy is not armed to respond to the severity of climate change. Opposite to the idea that environmental movements lead to democratization, it is based on science and ecological necessity, experts' knowledge, and a non-participatory approach to public policymaking (Swyngedouw 2010, 225). The approach is designed around highly technocratic and hierarchical policy processes, while individual liberties and the action of non-state actors stay limited (Gilley 2012).

The concept of *environmental authoritarianism* first appeared in the 19th century when several authors pointed to the dilemma faced by democracy: how to respond to the needs of the planet while dealing with selfish consumerism (Heilbroner 1991; Ophuls and Boyan 1977, 1992). Although the debate faded following the collapse of Communist regimes in the USSR and Eastern Europe, scholars have recently revived it. Dryzek and Dunleavy (2009, 262–263), for instance, draw on China's "efficient" response to environmental issues to assess what environmental politics could become in an authoritarian regime. As Gilley (2012, 299) contends, Chinese institutions' ability to manage participatory processes can become an advantage. Contrary to more "democratic" (e.g., Thailand) or authoritarian (e.g., Myanmar) states, *environmental authoritarianism* ensures a complementarity between top-down and bottom-up mechanisms.

Such analyses recognize the CPC's commitment to protecting the Environment as a clue that China's authoritarian political system is better armed than democratic settings to cope with environmental issues. Yet this growing appeal of an authoritarian response to the urgent climate crisis is not only to be found in academia. Throughout the years, as I introduced my research to friends, university colleagues, or students, arguments stating that "at least China is doing something about it", or "China has the power to make actual change" have become normal reactions. Readers of this book have probably heard or stated similar opinions. Most such reactions contend that the CCP's firm hand is better equipped to enforce much needed and rapid actions (although the recent situation with the COVID-19 pandemic may have changed such positions). The severity of environmental conditions has thus revived old debates over the need for coercive governance mechanisms in democratic settings.

In fact, the desirability of an authoritarian approach to environmentalism is a subject that is increasingly present in the literature (Blühdorn 2013). Shearman and Smith (2007), for instance, argue that liberal democracy is a fundamental problem for environmental destruction. According to the authors, authoritarianism is the key to dealing with climate change. In the same line of thought, British scientist James Lovelock stated that democracy "must be put on hold" to deal with environmental challenges ahead, calling for more authoritarianism and less egalitarianism.[30]

In *Climate Leviathan: A Political Theory of Our Planetary Future*, Wainwright and Mann (2018) take two variables—the world's economic structure (capitalist or non-capitalist) and the distribution of world power (planetary sovereignty or anti-planetary sovereignty)—to propose four theoretical frameworks. One model, which they term "Climate Mao", advances rapid, revolutionary, state-led transformation (2018, 39). Wainwright and Mann base their proposal on the premise that the state should have full power over who can emit carbon, use resources, or regulate waste—and how they do so. Hence, they look to the Chinese authoritarian model of environmentalism as an appealing way to respond to the failures of liberal democracies.

Past literature has mainly focused on assessing the efficacy and policy outcomes (pros and cons) of authoritarian forms of environmental politics (Gilley 2012; Eaton and Kostka 2014; Ahlers and Shen 2018; Povitkina 2018); analyzing its benefits and limits in the understanding of actual environmental policies (Lo 2015; Sheng et al. 2018); or, evaluating governments' persuasive and coercive strategies (Moore 2014). In most cases, studies either lament the inefficiency of Western democracy or praise authoritarianism as a better alternative. Scholarship lacks consensus on the efficiency of authoritarianism versus democracy, however. I believe that analyzing empirical evidence of the effects of these "green" authoritarian alternatives at the local level may help advance the discussion. Notably, by showing how authoritarian practices reshape the role of non-state actors and individuals and associated consequences (as the CCP's distrust of civil society hinders its ability to effectively detect urgent issues, as the spread of COVID-19 has shown).

Assessing environmentalism as a new ideology to power is crucial. Literature has considerably advanced in measuring how authoritarian environmental governance has become in China. Defined as *authoritarian environmentalism* by many scholars (Gilley 2012; Eaton and Kostka 2014; Moore 2014; Han 2018; Ahlers and Shen 2018; Li et al. 2019; Shen and Jiang 2021), studies gauge the efforts of the CCP in keeping control over all environmental policy decisions using top-down and non-participatory mechanisms. A re-centralization of state power and a reduction of local autonomy characterizes the Party's coercive top-down governmental tools, techniques, and technologies (Eaton and Kostka 2014; G. C. Chen and Lees 2018).

Yet few of this analyses assess how environmentalism can serve authoritarianism. Joining Li and Shapiro's arguments (2020), I presuppose *environmental authoritarianism* better reflects the need to gauge the ways in which the CCP is exploiting the Environment as a new form of political capital. Specifically, I direct my interest towards assessing the dynamics between authoritarian positions on the Environment and the restructuring of Chinese civil society's role in governance, ultimately observing their effects on the party-state's capacity for resilience. By resilience, I mean the capacity of the CCP to respond and adapt to environmental issues while retaining its leading role, function, and identity (Walker et al. 2004). The concept of *resilience* opens the way to new questions: (1) is *environmental authoritarianism* an end to achieve more sustainable outcomes or sustainable outcomes a tool to achieve regime stability?; (2) can participatory mechanisms minimize disruption and reinforce the authoritarian regime's stability?

"Green" participatory governance

To date, observations have suggested that authoritarian forms of environmental governance undermine local community and civil society through violent repression and dissent (Middeldorp and Le Billon 2019). Because policy processes follow command-and-control mechanisms and an autonomous state (Lo 2015), they embody the antithesis of more democratic and participatory forms of environmentalism (G. C. Chen and Lees 2018). In contrast to liberal political systems, participation by non-governmental actors serves the state and, thus does not encourage environmental politics (Beeson 2010; Gilley 2012).

Opposing earlier findings, in the following chapters, I present empirical evidence that several alternative forms of participatory governance, guided and facilitated by SGOs, have become part of urban environmental governance in contemporary China. My analysis of the situation in Shanghai proves that the state is purposefully blurring coercive mechanisms with processes of public participation. I claim governmental actors engage in civic forms of organization because it helps them to check and enforce policies. Even though some recognize environmental issues as a source of growing social unrest, especially in big cities (Economy 2004; Steinhardt and Wu 2016),[31] I argue that discussion of environmental issues can act as a powerful approach to instate grassroots party-building and, therefore, recentralize control. Thus, a growing implication of civil society in governance does not always signal a fundamental shift towards liberal democracy, rather the contrary: It can reinforce the Party's legitimacy.

This strategy opens new political spaces for new actors to emerge, however. Yet, as described in Figure 2.2, the state has reinforced coercive mechanisms to ensure all actors follow its plan. Sorted, graded, or dismissed, SGOs perform an important watchdog role under *environmental authoritarianism*. On the one hand, they verify whether local authorities pursue the agenda set at

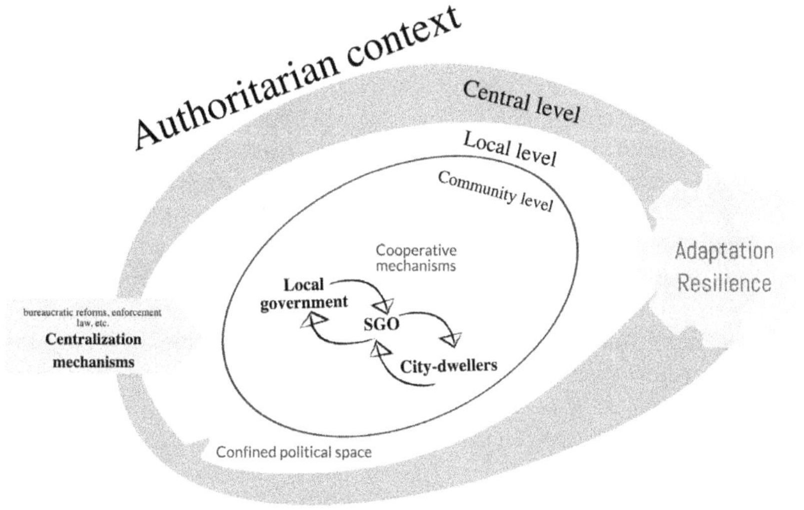

Figure 2.2. Environmental authoritarianism analytical framework.

the central level and, on the other, they help the local government to enforce environmental compliance and citizen engagement. Such strategies are observable in waste management practices, as I will show in Chapter 4, but they are also directed towards creating novel forms of "green" citizenship.

This book departs from the counterintuitive assumption that rising public participation mechanisms strengthen the Communist Party's legitimacy.[32] It therefore contributes to the discussion of several scholars who have been focusing on the power of economic performance (White 1986; Wintrobe 2000), Confucian meritocracy (Bell 2016), institutionalization (Nathan 2017), fragmentation (Cai 2008), collaboration (Y. Jing and Hu 2017), or nationalistic sentiments (Pun and Qiu 2020) to assess authoritarian resilience.

The phenomenon of "authoritarian modernity" is understood as a set of policies intended to achieve economic development, while demobilizing civil society beyond the modernization agenda (Thompson 2015; Gel'man 2016). Taking the case of China, Mark Thompson asserts that long-term claims to power have been characterized by a rejection of liberalism and democracy as "Western" and, thus, culturally alien (2015, 8), with the country's economic performance playing a leading role in this justification of power. Yet, as explored here above, environmental issues came to weaken the belief that the Chinese population was the grateful beneficiary of economic advancement. The discontentment of Chinese populations over environmental issues has brought many frustrations to the state's pragmatic legitimation.

In her book *Civil Society under Authoritarianism* (2014), Jessica C. Teets explains that government officials in China learned to partner with groups emerging from civil society. This model, which Teets coins "consultative authoritarianism", channels for citizens to act under the conditions of authoritarianism.

From Hu Jintao to the new President, Xi Jinping, however, she warns that civil society governance has evolved towards more repression and control. But does this mean that processes of "consultation" or "collaboration" have totally disappeared? In this book, I argue they have not, but they have become more sophisticated and institutionalized. The CCP has rearranged these processes to increase state control and avoid the dangers of citizen mobilization and protest.

Using *environmental authoritarianism* as an analytical framework enables us to assess how environmental protection is being used by authoritarian regimes to extend their legitimacy to power. Deconstructing these processes help us grasp how urban power is being exercised, reproduced, and adapted in China's "authoritarian capitalist" context (Sallai and Schnyder 2020). I will examine here how new power interactions between (1) the government and the private/social sector and (2) the government and citizens come together under the "ecological civilization" model.

Power beyond coercion

I will focus in particular on authoritarian power and practices beyond the logics of openness versus control or pluralism versus monism (Owen 2020). The empirical chapters sketch how adaptations to environmental issues intensify authoritarian practices by making participatory and coercive mechanisms coexist and going beyond normative conceptions of civil society under authoritarian rule. In fact, the CCP has recognized that, to deal with contemporary urban planning, it needs to rely on bottom-up participation.

This new necessity was embodied by a series of explosions at a storage container in Tianjin in 2015, which resulted in the death of over 160 people and injured hundreds of others. The event was symbolic of not only the pitfalls of rapidly urbanizing economies but above all, the danger of China's lack of enforcement capacity and transparency, of government corruption, and how this threatens the regime's legitimacy. Merely implementing top-down mechanisms, such as the "lifetime accountability" system for cadres or including green criteria into cadre evaluations, cannot prevent local officials from taking advantage of China's decentralized structure. Or, in Johnson's own words, the "target responsibility system is not a panacea for mitigating all environmental risks, and needs to be supplemented by greater supervision both within the system and from non-state actors" (2017, 203).

Recent research on civil society in authoritarian regimes has proven that civil society and non-state actors can contribute to legitimating authoritarian rule (Lorch and Bunk 2017; Fröhlich and Skokova 2020; Toepler et al. 2020). However, this finding has neither been used to explore the Chinese context and/or *environmental authoritarianism. China's Green Consensus* seeks to fill this gap by presenting an in-depth ethnographic analysis of the Shanghai municipality. By doing so, it shows that this inclination towards "cooperation" in environmental governance embodies the trend of authoritarian *post-politics* (Lam-Knott et al. 2019).

Hegemony in disguise

Hutter defines the inclusion of new actors in regulatory activities as "regulatory pluralism" (2017, 202). This inclusion represents a "responsive" form of authoritarianism aimed at dealing with the lows of China's political decentralization (van Rooij et al. 2016). The arguments advanced here go deeper into this question. Building on the concept of *post-politics*, I argue that this mode of governance advances a societal consensus, which is difficult to question or oppose, around the "common good" of environmental sustainability and the need for "good citizenship". To make Shanghai one of the most sustainable cities in the world, the political elite advances ambitious goals which, most of the time, lack specific content: making Shanghai the "global city", the "green city", or the "ecological civilization". Acting as "empty signifiers" (Swyngedouw 2011, 2016; Laclau 2017, 1996, 2005) these concepts mobilize consent and push multiple actors (SGOs and citizens) to engage within a hegemonic framework that is predetermined by those in power (Lam-Knott et al. 2019, 2).

In recent years, a vast body of research (mainly in urban political ecology) has explored the contradictions inherent to the current incarnation of sustainable urbanism (Adscheid and Schmitt 2021; Angelo and Wachsmuth 2020; Castán Broto 2020). Research from both the global North and South has shown that the use of sustainability as an "empty signifier" promotes the integration of diverse goals while reducing differences and possibilities for contestation (Brown 2016; Ernstson and Swyngedouw 2019). Yet despite a few exceptions (Lam-Knott et al. 2019), very few researchers have focused on the trend of *post-politics* in Asia. Notably, there has been little work on how cities and sustainability are being used to reinforce authoritarian legitimacy and restructure non-state actors' roles through consensual "green" narratives.

Mapping new post-political modes of urban governance with Actor-Network Theory

To investigate and trace the relations from which this "green" consensus has emerged, I used Actor-Network-Theory (ANT) as a methodological approach. As briefly outlined in the previous chapter, one of the major claims of ANT is that "political" issues result from a multitude of relations through which socio-material environments are co-created and enacted. ANT, thus, offers a means to develop a detailed analysis of the heterogeneous relations through which *environmental authoritarianism* tries to produce new "green" and "self-governed" citizens. ANT enables us in short to consider: how a collective *We* around sustainability is created, sustained and (re)negotiated beyond macro- and meso-level considerations (Anderson 2013 cited in Adscheid and Schmitt 2021).

Many readers may not be familiar with ANT, so it is useful here to take a simple example. Imagine you are working as a concierge in a hotel, and you have a mission: making sure guests do not lose their keys when leaving the hotel. This is your program. You start by leaving a notice on the back of the hotel room doors asking guests to kindly deposit their key at reception when leaving the hotel. However guests still continue to take their keys. They have an anti-program. What could you do to compel guests to change their behavior? After a few days of reflection, you decide to attach a weight to the key. Because the weight is heavy and troublesome, guests leave the key at the reception. Contrary to the notice on the door that simply requests that guests act in a certain way, the weight actually shapes the guests' behavior. It has agency. Outlined by Latour (1990), this simple example not only shows the role of non-humans in actor-networks, but also demonstrates how non-humans or objects can make a difference in the world.

Because ANT accounts for the tangled negotiations between human and non-human components—both referred to as *actants*—it is suitable for comprehending complex theoretical conceptualizations of agency and, therefore, helps us move beyond the political regime's dualities (authoritarian/democratic) (Ahlers 2018). When studying authoritarian settings, most studies concentrate on the role of the state, neglecting the intermediary role played by SGOs, individual actors, smog, or waste facilities. When SGOs become the focus, however, the analysis is based on expectations associated with civil society, as it is understood in Europe and North America (Salmenkari 2017). According to ANT, however, an *actant* is "anything that does modify a state of affairs by making a difference" (Latour 2005).

I use ANT as a methodological approach to crack open the *black box* of "ecological civilization" and examine its inner workings. To cite Callon and Latour (1981, 285), "A black box contains that which no longer needs to be reconsidered, those things whose contents have become a matter of indifference". *Black box* here can refer to almost anything: an algorithm, the Internet, an airplane, or the United Nations. To give you an idea, a computer is a *black box* because you understand what goes into it—sound cards, memory, motherboard—and what comes out—images, messages to your friends, videos—yet the process by which the inputs turn into outputs are a mystery for most users. When something runs efficiently, we focus only on its inputs and outputs, remaining ignorant of its internal complexity. Similarly, when talking about governance processes in China, we accept the CCP has complete control, yet we remain unaware of what makes its power possible.

This book does not see the CCP as a central and unified actor. It sees the state as an assemblage, "a phenomenon of *intracon-sistency*" (Deleuze and Guattari 1987, 478), an effect rather than the origin of power (Müller 2015, 32). It results from the associations established between a diversity of *actants* that make the difference as to whether one becomes more powerful than the other (*ibid.*). Of course, this statement directly contradicts the usual claims of oppositions between individuals and coercive states (*ibid.*, 30). According to ANT, once an actor-network has become irreversible, or *black boxed*, it becomes difficult for enrolled actors to escape the network, and so their behaviors become increasingly delimited. I make use of a network perspective in Chapters 4 and 5 to assess how actors are defined, associated, and simultaneously obliged to remain faithful to their alliances for environmental governance in contemporary China (Latour 2005, 6).

Because of some specific limits, ANT has provoked critical assessments, such as those raised by feminist scholars (Haraway 1988; Wajcman 2004). ANT's failure to recognize the distinctive properties of social structures, for instance, has been accused of weakening the theory's potential to explain both how social forces came about as well as their impact upon the world (Elder-Vass 2020). Still, by ethnographically engaging with ANT, I examine the multiple and coordinated practices through which the regime enacts *environmental authoritarianism.* ANT and its concepts were useful companions in describing how a "green" consensus—or post-political environment—is being built, mobilized, and upheld through heterogeneous relations between different *actants.*

To draw inferences about the distinctiveness of the causal capacities of the CCP and the roles that culture and social structure play in its capacities, I draw on Foucault's concept of *governmentality* and ethnographic observations. I will develop this point later. What is important to highlight for now is the fact that, despite limits, I found ANT useful in describing how *environmental authoritarianism* takes place because this approach allowed me to develop an analysis free from micro-macro distinctions (Müller 2015, 33), on the one hand, and to observe how a diversity range of what could be defined as "secondary" actors—human and non-human—contribute to the design of a common sense ideology, on the other.

In the broader literature, scholars study SGOs as extensions of the immutable and potent logic of the Chinese state (macro-perspective) or as a set of individuals working for their interests (micro-perspective). By putting the macro and micro on the same level, a network analysis better describes state-society interactions. It opens a path to assessing questions such as how SGOs negotiate their growing "niche" with the state (Akrich et al. 2006) to support themselves and/or to influence policy (Saich 2000). I will develop this point further in Chapter 4 where I assess how ZeroWaste, a leading organization in waste reduction in Shanghai, became a key actor in promoting community building through "cooperative" governance mechanisms with the state. In this context, ANT's sensitivity to the material world makes it an appropriate tool to analyze how varying rationalities—political, economic or social—contribute to the creation and spread of a post-political city. Chapter 5 will explore this point by focusing on how the activist collective Farming, by embracing market strategies, cultivates apolitical narratives that reaffirm the "empty signifiers" advanced by the political elite.

Creating consensus

According to ANT, when a few come to speak for the many, it is because they went through practices of *translation*. The concept of *translation* is a key concept of ANT, which is also known as the sociology of translation. Briefly, for Callon (1986), network building is about a leading actor who effectively *translates* his/her definition of a problem/approach/idea to other actors. Chapter 4 and 5 turn to the processes through which the "green" consensus is transformed into facts and artefacts, more specifically by applying Callon's four moments of *translation*: *problematization, interessement, enrollment* and *mobilization* (Guzman and De Souza 2018, 923). These four moments will tell the story of our case studies and describe how their dynamic works to enact particular realities and thus reveals how a "green" consensus is being *black boxed* outside the sphere of the state (Huang 1993). This analysis will allow us to understand the complex ordering of *environmental authoritarianism* and how the *translation* of ZeroWaste and Farming have come to promote the creation of post-political urban environments (assessed in detail in Chapter 6).

This book explores the heterogeneity of actors and their constantly (re) negotiating relationships in order to render the situated making, creating, and emerging of *environmental authoritarianism*. Applying ANT opens a more people-and practice-centered perspective because it looks at the actors' modes of production on site. As such, I pay attention to the role of the state but also to the material world around it, which enables its durability and agency. Taking this lens enables us to deconstruct politics beyond the content of representations and understand how authoritarianism is adapting to extra pressures and, more importantly, how bottom-up initiatives become necessary for the *translation* process as it creates consensus regarding differences.

According to Latour, innovation is a process of *translation* through which a vague initial idea is shaped, diverted, and consolidated to build up a network of allies who believe in, test, and carry forward the development of the innovation. I use ANT to liberate my analysis from a "naïve" unidirectional view of power whereby the dominated (society) can only obey or resist to the dominant (state). The four stages of *translation* are briefly presented in Figure 2.3 and the application of this model is further explored in Chapters 4 and 5 when presenting the cases.

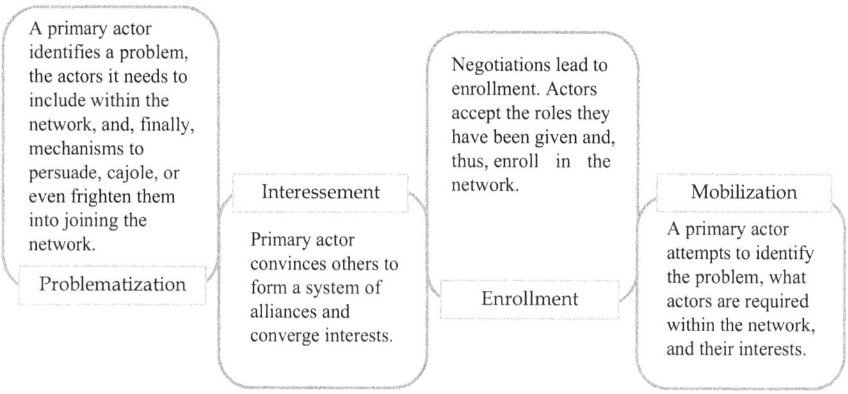

A primary actor identifies a problem, the actors it needs to include within the network, and, finally, mechanisms to persuade, cajole, or even frighten them into joining the network.
Problematization
Interessement
Primary actor convinces others to form a system of alliances and converge interests.
Negotiations lead to enrollment. Actors accept the roles they have been given and, thus, enroll in the network.
Enrollment
Mobilization
A primary actor attempts to identify the problem, what actors are required within the network, and their interests.

Figure 2.3. Actor-Network Theory: four moments of translation.

Conclusion

This chapter has discussed how China's "ecological civilization" has challenged theoretical assumptions that have previously been taken for granted, regarding the relationship between environmentalism and democracy. Even though the CCP's legitimacy has been questioned for its damaging environmental outcomes by civil society, several mechanisms and adaptation strategies were put in place to respond to emerging concerns. In this chapter, I develop the theoretical perspective of *environmental authoritarianism* and assert that China has been developing an environmental governance model based on both coercive and cooperative mechanisms. This chapter serves as an introductory section in our understanding of how the party-state is applying regulatory mechanisms of authoritarian co-optation to control China's third-sector development. Specifically, this chapter argues that the government has been promoting public trust by collaborating with SGOs, thereby fostering the image of a responsive and "green" government.

I argue that an *environmental authoritarianism* theoretical framework is useful for analyzing the mechanisms and processes which influence environmental governance in China and for explaining the CCP's capacity to respond to the threats of its severe environmental conditions. Accordingly, ANT theoretical resources have been introduced here and will be further developed in the empirical chapters to examine the development process of environmental governance and the adoption of aligned and contesting strategies. In Chapter 4, ANT will be applied to analyze how the CCP maintains its power through its ability to reach a great diversity of actors through the development of various networks and the use of specific "technologies". Then, in Chapter 5, I explore how some organizations engage in similar strategies to continue developing with a certain degree of freedom. The concepts from ANT are particularly relevant when exploring how actors take part in negotiation processes to achieve particular goals (e.g., when persuading SGOs to take part in the government agenda).

In sum, the theoretical frameworks of *environmental authoritarianism* and ANT are used in this book to: (1) make sense of the post-political context in which SGOs operate; and (2) show how the context increasingly limits non-state actors' ability to engage in contentious action. Shanghai, in particular, has developed various methods to keep SGOs' development under a "tight leash". In the following chapter, I explore the city's unique conditions and how it has managed to place SGOs' actions at the margins of authoritarianism, rather than challenging the status quo.

Notes

1 Source: Whale and Dolphin Conservation, available at https://uk.whales.org/whales-dolphins/species-guide/baiji/ (accessed 14 November 2019).

2 (2 January 2021) China Focus: Ten-Year fishing ban begins in Yangtze River. *Xinhua*. http://www.xinhuanet.com/english/2021-01/02/c_139635401.htm (accessed 7 April 2021).

3 USC US-China Institute (2008) *Air Quality at the 2008 Beijing Olympics.* University of Southern California. https://china.usc.edu/air-quality-2008-beijing-olympics (accessed 13 January 2017).
4 *Ibid.*
5 (2013) China acknowledges 'cancer village'. *BBC News.* http://www.bbc.com/news/world-asia-china-21545868 (accessed 14 April 2015); Jonathan Watts (2010) China's 'cancer villages' reveal dark side of economic boom, *The Guardian.* www.theguardian.com/environment/2010/jun/07/china-cancer-villages-industrial-pollution (accessed 12 November 2019).
6 See the Ministry of Ecology and Environment of the PRC (2018) "2017 Report on the State of the Ecology and Environment in China", available at http://english.mee.gov.cn/Resources/Reports/soe/SOEE2017/201808/P020180801597738742758.pdf (accessed 23 November 2019).
7 (2016) "2015 State of Environment Report Review", China Water Risk, available at http://www.chinawaterrisk.org/resources/analysis-reviews/2015-state-of-environment-report-review/ (accessed 23 November 2017).
8 Steven Mufson (2015) This documentary went viral in China. Then it was censored. It won't be forgotten, *The Washington Post.* www.washingtonpost.com/news/energy-environment/wp/2015/03/16/this-documentary-went-viral-in-china-then-it-was-censored-it-wont-be-forgotten/?noredirect=on&utm_term=.b24d5fa96270 (accessed 13 November 2018).
9 Semi-structured interview with an Associate Professor in Environmental Education and Education for Sustainability, 5 April 2016, Minhang, Shanghai.
10 See Hasmath and Hsu's book *NGO Governance and Management in China* (2015) for a detailed description of the diversity and complexity of the relationship between contemporary Chinese NGOs and the government.
11 Social organizations refer to membership associations (sectoral associations, academic associations, etc.), representing a specific social sector or a large social group. Civil non-enterprise units refer to public institutions, social groups, or social organizations established with the use of non-state-owned assets for the purpose of engaging in non-profit social service activities.
12 Christina Larson (2008) China's Emerging Environmental Movement. *YaleEnvironment360.* https://e360.yale.edu/features/chinas_emerging_environmental_movement (accessed 23 January 2016).
13 Congressional-Executive Commission on China (2015) "2015 Annual Report", *U.S. Government Publishing Office.* https://www.cecc.gov/sites/chinacommission.house.gov/files/2015%20Annual%20Report.pdf (accessed 15 January 2018).
14 Refer to Kendra Brock (2019) Environmental Protest Breaks out in China's Wuhan City. *The Diplomat.* https://thediplomat.com/2019/07/environmental-protest-breaks-out-in-chinas-wuhan-city/ (accessed 6 April 2021).
15 In the Report to the Seventeenth National Congress of the CPC (15 October 2007) Hu Jintao argued in favour of "expand[ing] people's democracy" and advocated the need to "expand the citizens' orderly participation in political affairs at each level and in every field". Source: *China Daily,* available at http://www.chinadaily.com.cn/china/2007-10/25/content_6204663.htm (accessed 24 May 2016).
16 Source: World Report 2015 China, available at https://www.hrw.org/world-report/2015/country-chapters/china-and-tibet (accessed 24 April 2017).
17 Source: Chinese Human Rights Defenders (2014) "A Nightmarish Year Under Xi Jinping's 'Chinese Dream'. 2013 Annual Report on the Situation of Human Rights Defenders in China", available at https://www.nchrd.org/wp-content/uploads/2014/03/FINAL-PDF_2013_CHRD-Report-on-Human-Rights-Defenders-compressed.pdf (accessed 16 October 2018).
18 I use "majority" because many Chinese citizens use Virtual Private Networks (VPN) to bypass China's firewall.

19 Eva Pils's presentation at the International Conference "Academic Freedom Under Threat", held at the Université Libre de Bruxelles on 11 December 2018.

20 Refer to "Human Rights Watch, World Report 2015: China", available at https://www.hrw.org/world-report/2015/country-chapters/china-and-tibet#eaa21f (accessed 23 November 2019).

21 Source: Qiao Long (2016) Chongqing 'luye xingdong' chuangban ren zao jingfang qiangxing 'he cha' 重庆"绿叶行动"创办人遭警方强行'喝茶' [The founder of Chongqing's 'Green Leaf Action' forced to 'drink tea' with police], *Radio Free Asia.* https://www.rfa.org/mandarin/yataibaodao/renquanfazhi/ql1-12232016101838.html (accessed 27 August 2017).

22 Source: Chinese Human Rights Defenders (31 May 2019) "Xue Renyi 薛仁义", available at https://www.nchrd.org/2019/05/xue-renyi/ (accessed 18 November 2019).

23 Source: Gao Feng (2019) Chongqing 'luye xingdong' chengyuan pan bin bei mimi panxing 重庆'绿叶行动'成员潘斌被秘密判刑 [Pan Bin, a member of 'Chongqing's Green Leaf Action' secretly sentenced], Radio Free Asia. https://www.rfa.org/mandarin/yataibaodao/renquanfazhi/gf2-01082019102613.html (accessed 23 December 2019).

24 Liu collected soil and paddy samples to examine the level of heavy metal pollution in the waters of Lake Dongting, a shallow lake in northeast Hunan.

25 Many of the arrests are reported on Front Line Defenders' website. See https://www.frontlinedefenders.org/en/location/china

26 (20 May 2018) Kaichuang meili zhongguo jianshe xin jumian 开创美丽中国建设新局面 [A new prospect for the creation of a Beautiful China], *XinhuaNews.* http://www.gov.cn/xinwen/2018-05/20/content_5292261.htm (accessed 6 January 2019).

27 Part of China's political discourse, the "Chinese Dream" has been promoted by Beijing's propaganda departments and relies on a "Chinese Way" which is "Socialism with Chinese characteristics". To achieve the "Dream", a citizen's dream must adhere to the common national goals which are dictated by the CCP. It is through the "Chinese Dream" that the Chinese nation will see a great rejuvenation. See Geremie R. Barmé (2013) "Chinese Dreams (中国梦)", *The China Story*, available at https://www.thechinastory.org/yearbooks/yearbook-2013/forum-dreams-and-power/chinese-dreams-zhongguo-meng-%E4%B8%AD%E5%9B%BD%E6%A2%A6/ (accessed 28 September 2019).

28 See Wen Ya (ed.) (21 September 2015) Zhonggong zhongyang guowuyuan yinfa 'shengtai wenming tizhi gaige zongti fang'an' 中共中央国务院印发'生态文明体制改革总体方案' [The Central Committee of the Communist Party of China issued the 'Overall Plan for the Reform of Ecological Civilisation System'], available at the State Council of the Republic of China website: http://www.gov.cn/guowuyuan/2015-09/21/content_2936327.htm (accessed 15 October 2018).

29 See Heidi Wang-Kaeding (6 March 2018) What Does Xi Jinping's New Phrase 'Ecological Civilization' Mean?. *The Diplomat.* https://thediplomat.com/2018/03/what-does-xi-jinpings-new-phrase-ecological-civilization-mean/ (accessed 22 May 2019).

30 See Leo Hickman (March 2010) James Lovelock on the Value of Sceptics and Why Copenhagen Was Doomed, *The Guardian.* https://www.theguardian.com/environment/blog/2010/mar/29/james-lovelock (accessed 13 May 2019).

31 Normally, environmental regulations are stricter in richer eastern seaboard areas. For instance, in some cases such as the 2007 PX protest in Xiamen, public resistance resulted in the polluting project being relocated to poorer areas.

32 Indeed, as Johnson (2017, 203) argues: "Although improving stakeholder accountability is a crucial aspect of enhancing China's environmental risk regime, doing so through empowering the public is perceived as being in conflict with CCP resilience that is increasingly associated with maintaining social stability".

References

Adscheid, Toni, and Peter Schmitt. 2021. "Mobilising Post-Political Environments: Tracing the Selective Geographies of Swedish Sustainable Urban Development". *Urban Research & Practice* 14 (2): 117–137.

Ahlers, Anna L. 2018. "Introduction: Chinese Governance in the Era of 'Top-Level Design'". *Journal of Chinese Governance* 3 (3): 263–267.

Ahlers, Anna L., and Yongdong Shen. 2018. "Breathe Easy? Local Nuances of Authoritarian Environmentalism in China's Battle against Air Pollution". *The China Quarterly* 234: 299–319.

Akrich, Madeleine, Michel Callon, and Bruno Latour. 2006. *Sociologie de La Traduction: Textes Fondateurs.* Presses Des Mines.

Angelo, Hillary, and David Wachsmuth. 2020. "Why Does Everyone Think Cities Can Save the Planet?" *Urban Studies* 57 (11): 2201–2221.

Battěk, Rudolf, and Paul Wilson. 1985. "Spiritual Values, Independent Initiatives and Politics". *International Journal of Politics* 15 (3/4): 97–109.

Beeson, Mark. 2010. "The Coming of Environmental Authoritarianism". *Environmental Politics* 19 (2): 276–294.

Bell, Daniel A. 2016. *The China Model: Political Meritocracy and the Limits of Democracy.* Princeton University Press.

Bina, Olivia. 2008. "Context and Systems: Thinking More Broadly About Effectiveness in Strategic Environmental Assessment in China". *Environmental Management* 42 (4): 717–733.

Birnie, Patricia, Alan Boyle, and Catherine Redgwell. 2009. *International Law and the Environment.* Oxford University Press.

Blühdorn, Ingolfur. 2011. "The Politics of Unsustainability: COP15, Post-Ecologism, and the Ecological Paradox". *Organization and Environment* 24 (1): 34–53.

Blühdorn, Ingolfur. 2013. "The Governance of Unsustainability: Ecology and Democracy after the Post-Democratic Turn". *Environmental Politics* 22 (1): 16–36.

Brehm, Stefan, and Jesper Svensson. 2017. "A Fragmented Environmental State? Analysing Spatial Compliance Patterns for the Case of Transparency Legislation in China". *Asia-Pacific Journal of Regional Science* 1 (2): 471–493.

Brock, Kendra. 2019. Environmental Protest Breaks out in China's Wuhan City. *The Diplomat.* https://thediplomat.com/2019/07/environmental-protest-breaks-out-in-chinas-wuhan-city.

Brown, Trent. 2016. "Sustainability as Empty Signifier: Its Rise, Fall, and Radical Potential". *Antipode* 48 (1): 115–133.

Cabestan, Jean-Pierre. 2004. "Is China Moving towards 'Enlightened' but Plutocratic Authoritarianism?" *China Perspectives* 2004 (55).

Cabestan, Jean-Pierre. 2014. *Le Système Politique Chinois.* Presses de Sciences Po.

Cai, Yongshun. 2008. "Power Structure and Regime Resilience: Contentious Politics in China". *British Journal of Political Science*, 411–432.

Callon, Michel. 1986. "Eléments Pour Une Sociologie de La Traduction. La Domestication Des Coquilles Saint-Jacques et Des Marins-Pêcheurs Dans La Baie de Saint-Brieuc". *L'année Sociologique* 36: 169–208.

Callon, Michel, and Bruno Latour. 1981. "Unscrewing the Big Leviathan: How Actors Macro-Structure Reality and How Sociologists Help Them to Do So". In *Advances in Social Theory and Methodology: Toward an Integration of Micro-and Macro-Sociologies*, edited by K. Knorr and A. Cicourel, 277–303. Routledge.

Carmin, JoAnn, and Adam Fagan. 2010. "Environmental Mobilisation and Organisations in Post-Socialist Europe and the Former Soviet Union". *Environmental Politics* 19 (5): 689–707.

Castán Broto, Vanesa. 2020. "Beyond Tabulated Utopias: Action and Contradiction in Urban Environments". *Urban Studies* 57 (11): 2371–2379.

Chao, Y E, and Jennifer Onyx. 2015. "Development Paths, Problems and Countermeasures of Chinese Civil Society Organizations". *Cosmopolitan Civil Societies: An Interdisciplinary Journal* 7 (2): 1–18.

Chen, Geoffrey C., and Charles Lees. 2018. "The New, Green, Urbanization in China: Between Authoritarian Environmentalism and Decentralization". *Chinese Political Science Review* 3 (2): 212–231.

Chen, Jie. 2010. "Transnational Environmental Movement: Impacts on the Green Civil Society in China". *Journal of Contemporary China* 19 (65): 503–523.

Chen, Yuyu, Avraham Ebenstein, Michael Greenstone, and Hongbin Li. 2013. "Evidence on the Impact of Sustained Exposure to Air Pollution on Life Expectancy from China's Huai River Policy". *Proceedings of the National Academy of Sciences* 110 (32): 12936–12941.

Cooper, Caroline M. 2006. "'This Is Our Way in': The Civil Society of Environmental NGOs in South-West China". *Government and Opposition* 41 (1): 109–136.

Corry, Olaf. 2013. "The Green Legacy of 1989: Revolutions, Environmentalism and the Global Age". *Political Studies* 62 (2): 309–325.

Corsetti, Gabriel. 2019. "How Many NGOs Are There Really in China?" *China Development Brief* 2019.

Deleuze, Gilles and Felix Guattari. 1987. *A Thousand Plateaus. Capitalism and Schizophrenia*. University of Minnesota Press.

Dewalt, K. M., and B. R. Dewalt. 2001. *Participant Observation: A Guide for Fieldworkers.* AltaMira Press.

Dryzek, John S., and Patrick Dunleavy. 2009. *Theories of the Democratic State.* Palgrave Macmillan.

Eaton, Sarah, and Genia Kostka. 2014. "Authoritarian Environmentalism Undermined? Local Leaders' Time Horizons and Environmental Policy Implementation in China". *The China Quarterly* 218 (1): 359–380.

Eckersley, Robyn. 2020. "Ecological Democracy and the Rise and Decline of Liberal Democracy: Looking Back, Looking Forward". *Environmental Politics* 29 (2): 214–234.

Economy, Elizabeth C. 2004. *The River Runs Black: The Environmental Challenge to China's Future.* Cornell University Press.

Economy, Elizabeth. 2018. *The Third Revolution: Xi Jinping and the New Chinese State.* Oxford University Press.

Elder-Vass, Dave. 2020. *Actor-Network Theory.* SAGE Publications Ltd.

Ernstson, Henrik, and Erik Swyngedouw. 2019. *Urban Political Ecology in the Antropo-Obscebe: Interruptions and Possibilities.* Routledge.

Fei, Sheng. 2016. "Environmental Non-Government Organizations in China since the 1970s". In *Environment, Modernization and Development in East Asia*, edited by Liu Tj, and J. Beattie, 203–222. Springer.

Florence, Éric. 2014. "China, 1978–2013: From One Plenum to Another. Reflections on Hopes and Constraints for Reform in the Xi Jingping Era". *Madariaga Paper* 7 (7) (July).

Florence, Éric, and Pierre Defraigne. 2013. *Towards a New Development Paradigm in Twenty-First Century China: Economy, Society and Politics.* Vol. 12. Routledge.

Fröhlich, Christian, and Yulia Skokova. 2020. "Two for One: Public Welfare and Regime Legitimacy through State Funding for CSOs in Russia". *VOLUNTAS: International Journal of Voluntary and Nonprofit Organizations* 31 (4): 698–709.

Fu, Diana, and Greg Distelhorst. 2018. "Grassroots Participation and Repression under Hu Jintao and Xi Jinping". *The China Journal* 79 (1): 100–122.

Fulda, Andreas. 2017. "The Contested Role of Foreign and Domestic Foundations in the PRC: Policies, Positions, Paradigms, Power". *Journal of the British Association for Chinese Studies* 7. http://bacsuk.org.uk/wp-content/uploads/2017/07/JBACS-7-Fulda-p-63-99.pdf.

Geall, Sam. 2013. *China and the Environment: The Green Revolution*. Zed Books Ltd.

Gel'man, Vladimir. 2016. *Authoritarian Modernization in Russia: Ideas, Institutions, and Policies*. Taylor & Francis.

Gilley, Bruce. 2012. "Authoritarian Environmentalism and China's Response to Climate Change". *Environmental Politics* 21 (2): 287–307.

Guzman, Gustavo, and Mariana Mayumi P. De Souza. 2018. "Shifting Modes of Governing Municipal Waste—A Sociology of Translation Approach". *Environment and Planning A* 50 (4): 922–938.

Han, Heejin. 2018. "Legal Governance of NGOs in China under Xi Jinping: Reinforcing Divide and Rule". *Asian Journal of Political Science* 26 (3): 390–409.

Hansen, Mette Halskov, and Zhaohui Liu. 2018. "Air Pollution and Grassroots Echoes of 'Ecological Civilization' in Rural China". *The China Quarterly* 234: 320–339.

Haraway, Donna J. 1988. "Situated Knowledges: The Science Question in Feminism and the Privilege of Partial Perspective". *Feminist Studies* 14 (3): 575–599.

Hasan, Samiul. 2015. "Disseminating Asia's Third Sector Research". *Voluntas: International Journal of Voluntary and Nonprofit Organizations* 26: 1007–1015.

Heilbroner, Robert L. 1991. *An Inquiry into the Human Prospect: Looked at Again for the 1990s*. W. W. Norton & Company.

Hicks, Barbara. 1996. *Environmental Politics in Poland: A Social Movement Between Regime and Opposition*. Columbia University Press.

Ho, Ming-sho. 2011. "Environmental Movement in Democratizing Taiwan (1980–2004): A Political Opportunity Structure Perspective". In *East Asian Social Movements*, edited by Jeffrey Broadbent and Vicky Brockman, 283–314. Springer.

Ho, Peter. 2007. "Embedded Activism and Political Change in a Semiauthoritarian Context". *China Information* 21 (2): 187–209.

Hsu, Jennifer Y. J. 2014. "Chinese Non-Governmental Organisations and Civil Society: A Review of the Literature". *Geography Compass* 8 (2): 98–110.

Hsu, Jennifer Y. J., and Reza Hasmath. 2015. *NGO Governance and Management in China*. Routledge.

Hsu, Jennifer Y. J., and Reza Hasmath. 2017. "A Maturing Civil Society in China? The Role of Knowledge and Professionalization in the Development of NGOs". *China Information* 31 (1): 22–42.

Huang, Philip. 1993. "'Public Sphere'/'Civil Society' in China?: The Third Realm between State and Society". *Modern China* 19 (2): 216–240.

Hutter, Bridget M. 2017. *Risk, Resilience, Inequality and Environmental Law*. Edward Elgar Publishing.

Imbach, Jessica. 2020. "Is Green the New Red: Cultural Perspectives on Ecological Civilization". *Eurics* 5.

Jacobs, Jamie Elizabeth. 2002. "Community Participation, the Environment, and Democracy: Brazil in Comparative Perspective". *Latin America Politics and Society* 44 (4): 59–88.

Jie, Chen. 2006. "The NGO Community in China. Expanding Linkages With Transnational Civil Society and Their Democratic Implications". *China Perspectives* 2006 (68): 29–40.

Jing, Jun. 2000. "Environmental Protests in Rural China". In *Chinese Society: Change, Conflict and Resistance*, edited by Elizabeth J. Perry and Mark Selden, 143–160. Routledge.

Jing, Yijia, and Yefei Hu. 2017. "From Service Contracting To Collaborative Governance: Evolution of Government-Nonprofit Relations". *Public Administration and Development* 37 (3): 191–202.

Johnson, Thomas. 2013. "The Health Factor in Anti-Waste Incinerator Campaigns in Beijing and Guangzhou". *The China Quarterly* 214, 356–375.

Johnson, Thomas. 2017. "Environmental Risks and Authoritarian Resilience in China". In *Risk, Resilience, Inequality and Environmental Law*, edited by Bridget M. Hutter, 188–204. Edward Elgar Publishing,

Johnson, Thomas, Anna Lora-Wainwright, and Jixia Lu. 2018. "The Quest for Environmental Justice in China: Citizen Participation and the Rural–Urban Network against Panguanying's Waste Incinerator". *Sustainability Science* 13 (3): 733–746.

Kan, Haidong. 2009. *Environment and Health in China: Challenges and Opportunities.* National Institute of Environmental Health Sciences.

Kassiola, Joel Jay, and Sujian Guo. 2010. *China's Environmental Crisis: Domestic and Global Political Impacts and Responses.* Palgrave Macmillan.

Kim, Sunhyuk. 2000. "Democratization and Environmentalism: South Korea and Taiwan in Comparative Perspective". *African and Asian Studies* 35 (3): 287–302.

Kong, Deyu, Egil Ytrehus, Anne Jarmot Hvatum, and He Lin. 2014. "Survey on Environmental Awareness of Shanghai College Students". *Environmental Science and Pollution Research* 21 (23): 13672–13683.

Kornreich, Yoel. 2016. "Unorthodox Approaches to Public Participation in Authoritarian Regimes: The Making of China's Recent Health Care Reforms". In *Chinese Politics as Fragmented Authoritarianism*, edited by Kjeld Erik Brødsgaard, 89–109. Routledge.

Kostka, Genia, and Chunman Zhang. 2018. "Tightening the Grip: Environmental Governance under Xi Jinping". *Environmental Politics* 27 (5): 769–781.

Laclau, Ernesto. 1996. "The Death and Resurrection of the Theory of Ideology". *Journal of Political Ideologies* 1 (3): 201–220.

Laclau, Ernesto. 2005. "Populism: What's in a Name?" In *Populism and the Mirror of Democracy*, edited by Francisco Panizza. Verso.

Laclau, Ernesto. 2017. "Why Do Empty Signifiers Matter in Politics?" In *Deconstruction*, edited by Martin McQuillan, 405–413. Routledge.

Lam-Knott, Sonia, Creighton Connolly, and Kong Chong Ho. 2019. *Post-Politics and Civil Society in Asian Cities: Spaces of Depoliticisation*. Routledge.

Latour, Bruno. 1990. "Technology Is Society Made Durable". *The Sociological Review* 38 (1_suppl): 103–131.

Latour, Bruno. 2005. *Reassembling the Social: An Introduction to Actor-Network Theory.* Oxford University Press.

Leutert, Wendy. 2018. "Firm Control: Governing the State-Owned Economy Under Xi Jinping". *China Perspectives* 2018 (1–2): 27–36.

Li, Xiaoliang, Xiaojin Yang, Qi Wei, and Bing Zhang. 2019. "Authoritarian Environmentalism and Environmental Policy Implementation in China". *Resources, Conservation and Recycling* 145: 86–93.
Lieberthal, Kenneth, and Michel Oksenberg. 1988. *Policy Making in China: Leaders, Structures, and Processes.* Princeton University Press.
Lo, Kevin. 2015. "How Authoritarian Is the Environmental Governance of China?" *Environmental Science and Policy* 54: 152–159.
Lorch, Jasmin, and Bettina Bunk. 2017. "Using Civil Society as an Authoritarian Legitimation Strategy: Algeria and Mozambique in Comparative Perspective". *Democratization* 24 (6): 987–1005.
Lu, Yiyi. 2007. "Environmental Civil Society and Governance in China". *International Journal of Environmental Studies* 64 (1): 59–69.
Ma, Qiusha. 2005. *Non-Governmental Organizations in Contemporary China: Paving the Way to Civil Society?* Routledge.
Mah, Kate. 2017. "The Silent Gatekeeper: Authoritarianism and Civil Society in China". *Political Science Undergraduate Review* 2 (2): 66–72.
Marcuse, Gary. 2011. Waking the Green Tiger: Rise of a Green Movement in China. Film. Face to Face Media.
Marinelli, Maurizio. 2018. "How to Build a 'Beautiful China' in the Anthropocene. The Political Discourse and the Intellectual Debate on Ecological Civilization". *Journal of Chinese Political Science* 23 (3): 365–386.
Mertha, Andrew C. 2011. *China's Water Warriors.* Cornell University Press.
Middeldorp, Nick, and Philippe Le Billon. 2019. "Deadly Environmental Governance: Authoritarianism, Eco-Populism, and the Repression of Environmental and Land Defenders". *Annals of the American Association of Geographers* 109 (2): 324–337.
Moore, Scott M. 2014. "Modernisation, Authoritarianism, and the Environment: The Politics of China's South–North Water Transfer Project". *Environmental Politics* 23 (6): 947–964.
Müller, Martin. 2015. "Assemblages and Actor-Networks: Rethinking Socio-Material Power, Politics and Space". *Geography Compass* 9 (1): 27–41.
Nathan, Andrew J. 2017. "China's Changing of the Guard: Authoritarian Resilience". In *Critical Readings on Communist Party of China*, 86–99. Brill.
Ophuls, William, and Stephen A. Boyan. 1977. *Ecology and the Politics of Scarcity: Prologue to a Political Theory of the Steady State.* W. H. Freeman & Co.
Ophuls, William, and Stephen A. Boyan. 1992. *Ecology and the Politics of Scarcity Revisited: The Unravelling of the American Dream. Trends in Ecology & Evolution.* W. H. Freeman & Co.
Owen, Catherine. 2020. "Participatory Authoritarianism: From Bureaucratic Transformation to Civic Participation in Russia and China". *Review of International Studies* 46 (4): 415–434.
Povitkina, Marina. 2018. "The Limits of Democracy in Tackling Climate Change". *Environmental Politics* 27 (3): 411–432.
Pun, Ngai, and Jack Qiu. 2020. "'Emotional Authoritarianism': State, Education and the Mobile Working-Class Subjects". *Mobilities* 15 (4): 620–634.
Rohde, Robert A, and Richard AMuller. 2015. "Air Pollution in China: Mapping of Concentrations and Sources". *PloS One* 10 (8): e0135749.
Saich, Tony. 2000. "Negotiating the State: The Development of Social Organizations in China". *The China Quarterly* 161 (February): 124–141.

Sallai, Dorottya, and Gerhard Schnyder. 2020. "What Is 'Authoritarian' about Authoritarian Capitalism? The Dual Erosion of the Private–Public Divide in State-Dominated Business Systems". *Business & Society* 60 (6):1312–1348.

Salmenkari, Taru. 2017. "Conceptual Confusion in the Research on Chinese Civil Society". *Made in China* 2 (1).

Shapiro, Judith. 2013. "The Evolving Tactics of China's Green Movement". *Current History* 112 (755): 224.

Shearman, David J. C., and Joseph Wayne Smith. 2007. *The Climate Change Challenge and the Failure of Democracy*. Praeger.

Shen, Wei, and Dong Jiang. 2021. "Making Authoritarian Environmentalism Accountable? Understanding China's New Reforms on Environmental Governance". *The Journal of Environment & Development* 30 (1): 41–67.

Sheng, Jichuan, Michael Webber, and Xiao Han. 2018. "Governmentality within China's South-North Water Transfer Project: Tournaments, Markets and Water Pollution". *Journal of Environmental Policy and Planning* 20 (4): 533–549.

Shieh, Shawn. 2018a. "Remaking China's Civil Society in the Xi Jinping Era". *ChinaFile*. http://www.chinafile.com/reporting-opinion/viewpoint/remaking-chinas-civil-society-xi-jinping-era.

Shieh, Shawn. 2018b. "The Chinese State and Overseas NGOs: From Regulatory Ambiguity to the Overseas NGO Law". In *Nonprofit Policy Forum*. Vol. 9. De Gruyter.

Snape, Holly. 2021. "Cultivate Aridity and Deprive Them of Air: Altering the Approach to Non-State-Approved Social Organisations". *Made in China* 6 (1): 54–59.

State Council of the People's Republic of China. 2019. "Xi Jinping: Tuidong woguo shengtai wenming jianshe mai shang xin taijie 习近平：推动我国生态文明建设迈上新台阶 [Xi Jinping: The Construction of an Ecological Civilisation to a New Level.". http://www.gov.cn/xinwen/2019-01/31/content_5362836.htm.

Stec, Stephen. 2005. "'Aarhus Environmental Rights' in Eastern Europe". *Yearbook of European Environmental Law*. Vol. 5.

Steger, Tamara Shevaun. 2004. *Environmentalism and Democracy in Hungary and Latvia*. Syracuse University Press.

Steinhardt, H. Christoph, and Fengshi Wu. 2016. "In the Name of the Public: Environmental Protest and the Changing Landscape of Popular Contention in China". *China Journal* 75 (1): 61–82.

Swyngedouw, Erik. 2010. "Apocalypse Forever?: Post-Political Populism and the Spectre of Climate Change". *Theory, Culture and Society* 27 (2): 213–232.

Swyngedouw, Erik. 2011. "Whose Environment? The End of Nature, Climate Change and the Process of Post-Politicization". *Ambiente & Sociedade* 14 (2): 69–87.

Swyngedouw, Erik. 2016. "Trouble with Nature: 'Ecology as the New Opium for the Masses'". In *The Ashgate Research Companion to Planning Theory*, 317–336. Routledge.

Tager, James, K. Bass, Glenn and Lopez, Summer. 2018. *Forbidden Feeds: Government Controls on Social Media in China. Technical report*. Pen America.

Tai, John W. 2015. *Building Civil Society in Authoritarian China*. Vol. 20. Springer eBooks.

Tang, Shui-Yan, and Xueyong Zhan. 2008. "Civic Environmental NGOs, Civil Society, and Democratisation in China". *Journal of Development Studies* 44 (3): 425–448.

Teets, Jessica C. 2014. *Civil Society under Authoritarianism: The China Model*. Cambridge University Press.

Thompson, Mark R. 2015. *Authoritarian Modernism in East Asia*. Springer.

Toepler, Stefan, Annette Zimmer, Christian Fröhlich, and Katharina Obuch. 2020. "The Changing Space for NGOs: Civil Society in Authoritarian and Hybrid Regimes". *VOLUNTAS: International Journal of Voluntary and Nonprofit Organizations* 31 (4): 649–662.

Tong, Yanqi. 2005. "Environmental Movements in Transitional Societies: A Comparative Study of Taiwan and China". *Comparative Politics*: 167–188.

van Rooij, Benjamin, Rachel E. Stern, and Kathinka Fürst. 2016. "The Authoritarian Logic of Regulatory Pluralism: Understanding China's New Environmental Actors". *Regulation and Governance* 10 (1): 3–13.

Wainwright, Joel, and Geoff Mann. 2018. *Climate Leviathan: A Political Theory of Our Planetary Future*. Verso.

Wajcman, Judy. 2004. *TechnoFeminism*. Polity Press.

Walker, Brian, Crawford S. Holling, Stephen Carpenter, and Ann Kinzig. 2004. "Resilience, Adaptability and Transformability in Social–Ecological Systems". *Ecology and Society* 9 (2).

Wang, Fangfei. 2021. "Narrating China: Reading Li Ziqi and Fangfang from a Nationalist Perspective". Doctoral dissertation, Duke University.

Wang, Ming, ed. 2010. *Shehui zuzhi gailun* 社会组织概论 [An Introduction to Social Organisations]. Zhongguo shehui kexue chuban she 中国社会科学出版社 [China Social Sciences Press].

White, Stephen. 1986. "Economic Performance and Communist Legitimacy". *World Politics: A Quarterly Journal of International Relations* 38 (36): 462–482.

Wike, Richard, and Bridget Parker. 2015. *Corruption, Pollution, Inequality Are Top Concerns in China: Many Worry about Threats to Traditions and Culture*. Pew Research Center.

Wintrobe, Ronald. 2000. *The Political Economy of Dictatorship*. Cambridge University Press.

Xie, Lei, and Hein-Anton Van Der Heijden. 2010. "Environmental Movements and Political Opportunities: The Case of China". *Social Movement Studies* 9 (1): 51–68.

Xie, Lei, and Peter Ho. 2009. "Urban Environmentalism and Activists′ Networks in China: The Cases of Xiangfan and Shanghai". *Conservation and Society* 6 (2): 141.

Yang, Guobin, and Craig Calhoun. 2007. "Media, Civil Society, and the Rise of a Green Public Sphere in China". *China Information* 21 (2): 211–236.

Yang, Zhenjie. 2013. "'Fragmented Authoritarianism'—the Facilitator behind the Chinese Reform Miracle: A Case Study in Central China". *China Journal of Social Work* 6 (1): 4–13.

Yuen, Samson. 2014. "Disciplining the Party: Xi Jinping's Anti-Corruption Campaign and Its Limits". *China Perspectives* 3 (December): 41–47.

Zhan, Xueyong, and Shui Yan Tang. 2013. "Political Opportunities, Resource Constraints and Policy Advocacy of Environmental Ngos in China". *Public Administration* 91 (2): 381–399.

3 The cooperative road towards sustainability in Shanghai

Introduction

Inaugurated in September 2015, the eco-friendly Shanghai Tower (*Shanghai zhongxin dasha* 上海中心大厦) in Lujiazui, Pudong, is the world's second-tallest building after Dubai's Burj Khalifa. This monumental building added China to the long list of countries with the tallest skyscraper or fastest elevator in the world, but, more importantly, the tower is a prime example of green architecture to be emulated in the future. On the city's iconic skyline, at the top of the 632-meter tower, 200 wind turbines generate around 10 percent of the building's electricity. The tower also boasts 43 different sustainable technologies that reduce its total energy consumption. These technological advancements positioned the tower on the list of the most advanced sustainable tall buildings in the world, winning China several international awards and honors.

This building encapsulates the growing ubiquitous attention that the CCP pays to urban "sustainability". And, in a race for international recognition, Shanghai, more than any other Chinese city, takes a major role in promoting the path to Chinese "ecological civilization" (as mentioned in Chapter 2). As a poster child for China's economic miracle, the coastal Metropole epitomizes the country's exponential growth since the beginning of the reform era (post-1978) and openness to the world. The city's unique political, economic, and social conditions make it a good and interesting ground to understand China's green ambitions and consider how urban governance is being reshaped to respond to pressing social and environmental issues, and the ways in which Shanghai has been adapting environmental governance to reach its goals is the major theme of this chapter.

I had previously advanced that the politicization of environmental protection by the party and the state serves as a new ideology to power. Focusing on the theoretical framework of *environmental authoritarianism*, I advance the need for a new analytical lens to assess how this "green" consensus is creating unused spaces of encounter in which the state and society's views come together. While previous chapters sketched the big lines of *environmental authoritarianism*—with a focus on the central level and the party's efforts to (re)centralize power—I will

DOI: 10.4324/9781003231325-3

now shift the analysis towards the local level (see Figure 2.2). My aim is to further describe the political nature of the CCP's governance innovations, considering how these unfold within the city's goal of reaching sustainable goals, and why they characterize a cooperative rather than a collaborative model of governance.

Several scholars, such as Yijia Jing (2010; 2012; Jing and Hu 2017), have stated that Shanghai combines a strong tradition of social control and enthusiasm for engaging non-state actors in governance issues. Taking stock of the literature on collaborative governance (Ansell and Gash 2008; Teets and Jagusztyn 2015; Hou and Liu 2017; Jing and Hu 2017; Ratigan and Teets 2019), this chapter brings attention to new political practices and cooperative dynamics in the urban sphere. I focus on how, in order to respond to various challenges, local officials in Shanghai became frontrunners in regulating—rather than facilitating—the provision of essential social services. The city's appetite to become an international modern city, in particular, led it to become a pioneer in the purchase of social services from SGOs before the establishment of any national policy.

To explore this point, I go beyond the theoretical assumptions discussed in Chapter 2. In particular, I clarify how cooperative approaches rather than collaborative ones are being implemented and designed at a local municipal level. To do this, I first briefly discuss Shanghai's political commitment to making sustainable development a long-term development goal and give a short outline of the city's historical and conceptual context. I then explore the specificities of Shanghai's current environmental governance strategies and their correlation with the evolution of non-state actors in the local state's logic, considering: Why has the use of SGOs in delivering public services become key in Shanghai? What kinds of strategies did the Metropole develop to change its citizens' behavior (their role in society) without taking an environmentally totalitarian approach? Finally, I argue that "cooperative" mechanisms of this type are leading SGOs to be trapped within roles that perpetuate the state's vision of "ecological civilization".

Shanghai: a brief historical background

> Never was Shanghai's allure greater than at the beginning of the 1930s. Hong Kong's views might be more magnificent, Peking's monuments more ancient, Yokohama's climate more salubrious, and Singapore less expensive. Perhaps the natives in the provinces were more cordial, the accents in Soochow more pleasing, or the food better in Canton, but ask any Orient-bound traveler his prime destination and the answer would invariably be "Shanghai"!
>
> (Dong 2001)

Strategically located at the mouth of the Yangtze River, for over 150 years Shanghai has been both an economically and politically dynamic area, and was one of the first Chinese cities to formally open trade with Western countries in 1842. From the 1920s to the 1930s, as merchants and foreign settlers took control of the city, Shanghai experienced a golden age of global modernity.[1] During this period, Shanghai was known as "The Paris of the East, the New York of the West". In the wake of this booming environment, the city saw its industries, commerce and transportation grow rapidly. It was also during this period, in 1921, that Shanghai saw the birth of the Communist Party. From the 1930s, however, several episodes—such as raids and invasions by the Japanese, World War II, and the fights between Nationalists and Communists—put an end to this frenetic period of development. When Communists declared victory in 1949, Mao's centrally planned economy and later the Cultural Revolution further exacerbated the city's decline, and the party was over.

The Metropole entered a deep slumber, only rising from its ashes following the political and economic reforms of the 1980s and the establishment of Pudong (a district to the east of the city and the Huangpu River) as a Special Economic Zone in 1990. Yet by then years of economic and political tension had put Shanghai behind Singapore, Hong Kong, Taiwan, or even domestic cities like Shenzhen that all emerged as industrial powers around the same time (Chin 2016, 148). Worse, being China's number one industrial city had left a lasting and disastrous legacy for the city's environment (Weller 2006, 59). The Suzhou River and Huangpu River, known for their black color and foul odor, encapsulated the price Shanghai paid after years of chaotic economic development and urbanization (S. Zhao et al. 2006). To regain its leading position in Asia, Shanghai needed rapid change and strong government support.

This support came from social entrepreneurs and local leaders when they started pushing the idea that the city should become the country's economic center (Sheng 2019). Yet because Deng Xiaoping and conservative leaders were afraid that too many resources would be transferred to the market—since Shanghai was a strategic tax resource for the central government, accounting for no less than one-sixth of total national revenue (Y. Chen 2007, 231)—they rejected the suggestion. Following the Tiananmen Square Massacre and the international outrage it provoked, however, Chinese leaders changed their position: Shanghai would become the fresh face of China's reform to the outside world (*ibid.*). The megacity would subsequently take on a role as a symbol of China's modernization and development, and this changed the attention given to its image. Thus, it does not come as a surprise that the Metropole came to play such a leading role in showcasing the CCP's green ambitions.

Shanghai puts on a "green" face

Besides the tower described above, the Shanghai World Expo in 2010 epitomized the importance of the city in exposing the new "green" face of China. President Xi Jinping was a fierce supporter of the Expo. This was seen as the perfect opportunity to remedy the disastrous impression China had made to the world two years earlier, during the Olympic games (Chapter 2). The green construction of the Expo revealed the government's new emphasis on protecting and improving the environment. Even though Shanghai is not a typical Chinese city, it serves as a striking example of China's ambitions: economic growth, technological progress, urban focus, and global significance. As one of the world's most urbanized cities, it embodies the sustainability challenges cities need to address worldwide. Because it is an estuary megacity, Shanghai also needs to cope with challenges characteristic of climate change (e.g., rising sea levels) and, thus, can serve as a model for sustainable development for other cities in the future. The general secretary of the CCP is seemingly aware of Shanghai's potential, claiming during a grand gathering that celebrated the thirtieth anniversary of the development and opening of Shanghai's Pudong New Area in 2020, that Shanghai will impress the world and usher in a new era of building a modern socialist country.

Still, since the aforementioned Maoist period of isolation that focused on industrialization, it would be wrong to argue that Shanghai is only following the line of the party and the state. The megacity's unique position in the international scene has heavily influenced its development strategies (S. Chen 2017, 225). Contrary to other Chinese cities, international experts and preferential policies of the central government—especially since the opening of the Pudong New Area in 1990—gave the Metropole a good basis to integrate itself with international norms and to bargain for resources and autonomy from the central government (Chin 2016, 148). But this autonomy is not a panacea, as I explore below and in the following chapters.

Because economic growth has played a central role in the City's agenda, business interests have considerably jeopardized environmental protection, with efficiency and growth considerably affecting environmental conditions and inequalities (Fulong Wu 2010). A good deal of attention was paid to living conditions not because the municipality had a bigger budget but as a strategy for attracting investors and talent (Y. Chen 2007, 170). As Sheng (2019, 68) explains, "Green is also a relative concept; the powerful actors in the planning process decided what is green and what is not. They considered large open spaces as green, although creating it forced thousands of people to move". So, although Shanghai encapsulates the modern green face of China, it is important to note that the city has suffered from highly uneven economic growth, and economic interests prevailed over environmental ones. Vague laws and regulations, conflicts between different government departments, and central-local blame-shifting games have affected the development of better environmental implementation mechanisms (Ran 2017; Ran and Jian 2021).[2]

Tightening up environmental governance

Shanghai is one of four special municipalities in China with a status equal to a province. This means that the Shanghai Municipal Government reports directly to the central government in Beijing and follows nationwide guidelines. Yet Shanghai's contingent, fragmented, and heterogeneous nature has not always made for a smooth relationship with central level administration and supervision. A fuzzy allocation of responsibilities—termed *fragmented authoritarianism* in the literature (Lieberthal 1992; Mertha 2009; Yang 2013)—challenged the implementation of regulations set at a central level. As Brehm and Svensson (2017, 472) have pointed out, ample regulatory space between "vague" national laws and local implementation has enabled local actors to "hollow out" the principles of environmental regulation. As a result, it was not uncommon for economic development to be favored over environmental protection. There was a GDP-based appraisal system managed by lower levels of local government,[3] leading to an emphasis on economic growth, and a decentralized system of environmental governance which, in turn, created pollution and related problems rather than solving them (Brehm and Svensson 2017; Ran 2017). I explained previously that, to unify fragmented responsibilities, Xi's administration developed new bureaucratic reforms and enforcement mechanisms with the aim of (re)centralizing the considerable leeway enjoyed by local governments at the time (J. Wu et al. 2013).

Shanghai's role as a key hub, and its favorable connection to the outside world and access to foreign resources, proved a real challenge to the establishment and centralized governance. Yet the idea I want to convey here is that this new focus on "sustainability" was not only a result of Xi's efforts at centralization (see Chapter 2). Shanghai's environmental management went through different stages, as shown in Figure 3.1, and we can observe a turning point since the 2010s. In the rest of this book, I argue that this change results from three major factors: (1) harsher regulations coming from the central level highlighted in Chapter 2; (2) a rising middle-class organizing against

Figure 3.1. Major stages of Shanghai's urban environmental management.

vested economic interests (e.g., land-use rights) and pollution-related health problems (e.g., anti-paraxylene plant); and (3) a means of strengthening the city's legitimacy and competitiveness.

Environmental movements in Shanghai

The city's obsession with economic growth negatively affected its air, water, and soil but also considerably hindered the development of environmental movements. In the Metropole, environmental protests have been even more contained, and protests appeared relatively late when compared to other regions (Xie and Ho 2008). Apart from a few exceptions stemming from the student activism of the late 1980s (Lee 2007)—for example, the campaign against the construction of the New Jiangwan Town in 2001 (Xie and Ho 2008) or the 2015 anti-paraxylene protest in Jinshan[4]—a restrictive environment limited the space for environmental movements to emerge. In addition, the municipal government has largely constrained the development of SGOs in accordance with its own interests. Lee (2007) and Salmenkari (2017) advance other features that explain the docile, conservative, and self-contained nature of SGOs in Shanghai:

First, as WWF's Program Manager explained to me (interview 18 April 2016), there is a relatively small number of international SGOs active in the Metropole as compared to the capital where policies are made. When SGOs are present, however, they risk putting local actors at risk.[5] The media and scientific experts who collaborated with an international SGO during the Jiangwan Town case in 2001—the first collective protest to oppose environmental policy in Shanghai, for instance—were later intimidated by political authorities (Lie 2009).

As a second and related point, the city's position as a role model in China put the media under strict government surveillance, significantly limiting citizens' environmental awareness and capacity for mobilization (Lee 2007). Third, the municipal government's significant financial resources enabled it to deal with issues that might attract the attention of SGOs in other areas (e.g., migrant children). As an illustration, in 2017, investment in environmental protection accounted for 3.1 percent of the municipality's GDP (nearly 11 billion euros).[6] Such resources allow the state to co-opt SGOs, as I will explain later.

In addition to this rigid political context and a lack of favorable conditions for the nurturing of environmentalists, the institutionalization of social welfare and services into a corporatist model also slowed down the emergence and expansion of environmental movements.

Implementing a state corporatist model

In 1996, Mrs. Ma Yili 马伊里, then director-general of the social-development bureau in Pudong, identified an empty public facility in Luoshan Street that could be put to new use, perhaps as a kindergarten, she first thought. Because there were not enough children living in the area, she rapidly dropped the idea in favor of a citizens' club and started looking for a volunteer organization to manage it (Tuan et al. 2015). Mr. Wu Jianrong, secretary-general of the Shanghai Young Men's Christian Association, seized on the opportunity and submitted a proposal to the social-development bureau. After receiving the green light, the Luoshan Citizens' Club became the first case of a Chinese SGO being commissioned by a Chinese municipal government to deliver social services. By 2013, the Christian Association was managing nine centers which were constructed and subsidized by the Chinese municipal government (see Tuan et al. 2015).

Since then, this means of contracting services diffused and rapidly got the attention of the central government (Smith and Zhao 2016; Weng 2017). When I speak of purchasing, or "contracting out" (*goumai fuwu* 购买服务) services, I mean the practice of the government transferring public services to SGOs, enterprises and institutions through public bidding or targeted entrustment. In general, under a corporatist model, the government regulates social services rather than providing them. Because local governments implement the public services policies issued by the central government (welfare regionalism), they have shown enthusiasm for such corporatist arrangements (Leung and Xu 2015; Mok and Qian 2019).

These first experiments stimulated the National People's Congress to enact the first national law, the Government Procurement Law, aimed at regulating government procurement activities and enacted on 29 June 2002. Previously, government agencies had managed government procurement with the budgets of local governments.[7] From 2003, however, a uniformed law facilitated the creation of government procurement organizations, integrating new actors into China's state-dominated regulatory system while maintaining government guidance, and described by scholars as *state corporatism* (J. Y. J. J. Hsu and Hasmath, 2014; J. Han 2016; Jing 2018). Yet as Lei and Chak (2018) stress, the way in which the Chinese corporatist state unfolds at a local level varies from region to region and from service to service. As explored below, Shanghai has been a vanguard in creating favorable conditions for the development of service contracting. Although central government departments publish guidelines, local actors have made adaptations to facilitate regional variations and better respond to their social and economic needs.

Managed social innovation

Air pollution, the quality of public transport, or waste management have become important indicators for a city's competitive edge, meaning that finding rapid and effective solutions is a key priority for local leaders.[8] Moreover, as megacities such as Shenzhen keep expanding fast (J. Shen 2018), Shanghai is obliged to find solutions for environmental problems in order to keep its leading position as China's international economic center for trade, shipping, and finance. Public officials know that citizens have increasing expectations of high-quality public services, such as quality of schools, food safety, environmental protection, or health. Thus, the authorities need to maximize their response capacity to keep the city attractive to national and foreign investors (Liu 2015, 178).

According to Yijia Jing (2015b; see also Jing and Chen 2012; Jing and Hu 2017), expert on privatization, governance and collaborative service delivery at Fudan University, Shanghai leaders have been consciously introducing service-delivering into formal governance structures to respond to these pressures. The municipal government in particular has been developing multi-channel participation schemes and opening spaces for SGOs to take part in community governance and citizen-oriented processes and services (Jing and Gong 2012; Jing and Hu 2017).

Practitioners are constantly bringing this idea to the fore in conferences, and formal or informal interviews, depicting SGOs as central players in the enhancement of social innovation and the creation of new social interactions between the state and society. How these supposed horizontal connections prompted SGOs to assume roles and responsibilities that replace more traditional top-down command-and-control mechanisms is something I will continue to explore in the following sections.

The growth of SGOs in Shanghai: a prosperous but controlled environment

As the birthplace of the Chinese government purchase of services from SGOs (Weng 2017), Shanghai has been a testing ground since the Government Procurement Law came into effect on 1 January 2003 (Rothery 2003). That same year, the Shanghai Municipal Commission for Political Science and Law launched three service non-profit organizations—Xinhang, Yuangguang and Ziqiang—to prevent and reduce crimes in the city (Shi 2017). These organizations differ from the SGOs I observed in the field because the state created them to serve its own interests (scholars normally reference such organizations as Government-Organized Non-Governmental Organizations GONGOs). Still, these first trials opened the way to expanding new laws and regulations related to public welfare and charity in China.

On 10 June 2005, the State Council chose the Pudong New District of Shanghai to develop a pilot project on Comprehensive and Co-ordinated Reform, with the aim of giving more space to society and non-state organizations (Kerlin 2017). As a response, the Civil Affairs Office: (1) partially removed the system of dual registration of organizations; (2) allocated funding to grassroots projects; (3) created incubators to reinforce their development capacity; and (4) enhanced government procurement of services, especially to those organizations that were capable of providing for the unmet needs of local communities (D. Li 2016, 129).

It is important to note that many of these changes occurred under the leadership of Mrs. Ma Yili, mentioned above, who had already been a pioneer in government purchasing when she contracted Shanghai Young Men's Christian Association Hongkong to manage Luo Shan Community Centre in 1996. Head of the Shanghai Bureau of Civil Affairs from 2005 until 2013, she became a major player in the professionalization of SGOs in the Metropole. This constructive environment alongside Ma Yili's support encouraged private organizations such as the Non-Profit Incubator (NPI) (*en pai gongyi zuzhi fa zhan zhongxin* 恩派公益组织发展中心) to register without a government affiliation in Shanghai at the beginning of the 2000s (Fengshi Wu and Chan 2012).

Established in 2006 by Mr. Lü Zhao 吕朝, NPI has been promoting social innovation and cultivating social entrepreneurs by granting crucial support such as free or subsidized office space and supplies, IT support, mentoring, training or capacity building and, most importantly, a well-established network of government officials, business, and individual donors (Kolhoff 2016) to start-ups and small to medium-sized SGOs. From the beginning, the organization established a good, trust-based relationship with the government. It was actually Ma Yili herself who invited Lü to come to Shanghai and provided him with a "fast pass" to register NPI (D. Li 2016, 130) (the term "fast pass" was used by informant 7). Today, NPI has become one of the largest and most influential SGOs in China.[9] Satisfied with the incubator model, the Ministry of Civil Affairs has since promoted it nationwide (D. Li 2016, 132), and by 2020, there were over 30 NPIs throughout China.[10]

To further enhance the development of SGOs in the municipality, the local government launched other initiatives. In 2007, the Jing'an District established the Social Organization Association to provide services to non-profit organizations, such as non-profit incubation, registration assistance, capacity building, training, or information exchange. In 2012, the Jing'an District had a density of 14 organizations per 10,000 residents, compared to a rate of seven in Shanghai and three per 10,000 residents in China as a whole (Jing 2015b).

Then, in 2009, the Municipal Bureau of Civil Affairs of Shanghai started a program of venture philanthropy (*gongyi chuang tou* 公益创投) to facilitate young organizations' development, and chose NPI to operate the program. Since then, NPI leveraged over 200 million RMB to support over 600 organizations (D. Li 2016, 133). The Shanghai government also created a public welfare park—or Public Welfare Street—in Pudong District, with modern buildings to house SGOs that were sponsored by the local district. This Public Welfare Street was the first of its kind in Shanghai and China. Established in what used to be an old factory building, it housed around 15 SGOs at the time of my fieldwork. I entered the "incubator" on two occasions: first, to conduct a semi-structured interview with the Director of the BlueOcean SGO; and later, on one occasion as a volunteer. BlueOcean is an SGO that focuses on ocean issues and runs activities related to protecting the marine environment such as coastal clean-up activities or workshops in primary schools.

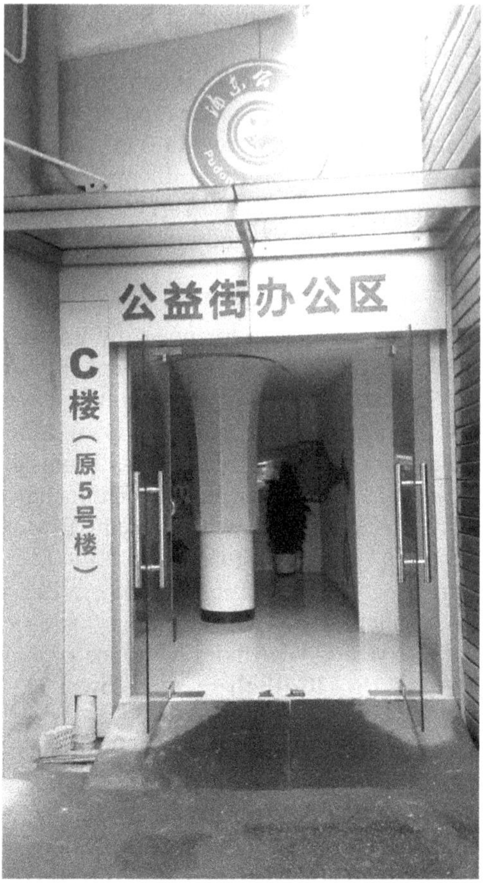

Figure 3.2. Entrance of Pudong Public Welfare Street.

Every time I entered the building, my passport identity was recorded, as was mandatory for foreigners, SGO members explained.

This "incubator" is part of Shanghai's "quick and controllable" environment, which, according to Jing (2015a, 601), has been enabled through excessive government intervention. Because SGOs remain under the control of one municipal government, they have less room to maneuver between different authorities (e.g., national, municipal or district level). Local governments mainly cooperate with organizations that are capable of covering gaps in support on issues where the state had given up or could not respond (e.g., waste, services for people with disabilities).

As a result, unlike Beijing and Yunnan, where SGOs try to reach beyond a single community, organizations in Shanghai are small and act locally. Several researchers have been studying the advancement and the impact of government procurement of public services at the local level (Fulong Wu 2002;

Rothery 2003; Jia and Su 2009; Jing and Gong 2012; Yu 2016; Gao and Tyson 2017). The central government took about ten years to developed mechanisms for government outsourcing, as showcased below (Table 3.1).[11] More recently, however, SGOs have faced increasing pressures to apply market strategies. In the following chapters, I will explore why the changes related to this business approach deserve attention. Yet, before delving into the on-the-ground reality of this growing commercialization and bureaucratization of SGOs in more detail, I will first clarify why these new relationships between the state and SGOs cannot be understood as "collaborative" governance. My aim here is to advance reflections on the subtleties of these close partnerships between the government and SGOs in China.

Table 3.1. Development phases of China's purchase of services. Sources: Jia and Su 2009; Carrillo et al. 2017; Yu and Guo 2019.

Stage 1—from 1996 to 1998 Research on government procurement of public services is put on the agenda in 1996. The Shanghai Civil Affairs Bureau started its pilot projects of government procurement
Stage 2—from July 1998 to June 2000 Local experimentations (e.g., Shenzhen) and scaling-up of pilot projects. Until 2000, most local governments announced regulations on government procurement; the government procurement law was on legislative agenda of the National People's Congress; relevant organizations of government procurement were established and improved; the scope and scale of government procurement increased at a rapid rate. By 2011, Shanghai had allocated ten % of its service expenditure to a competitive bidding process.
Stage 3—from June 2000 to December 2002 Scaling up of procurement, recognized nationally. On 29 June 2002, the National People's Congress approved Government Procurement Law of People's Republic of PRC.
Stage 4—from January 2003 to 2007 Developing stage to normalization and legalization. Based on the enlargement of government procurement, PRC's Ministry of Finance started the procedure of applying to be one party of the Agreement of Government Procurement in WTO in 2007.
Stage 5—from 2013 till 2015 In 2013, China announces new opportunities for the government's purchase of services with the *Guidance of the General Office of the State Council for Government Procurement of Public Service from Social Entities.* Several other documents were promulgated by multiple central ministries and commissions (Ministry of Finance, Ministry of Civil Affairs, etc.). After that, many local governments have issued related guidelines.
Stage 6—from 2016 to present Introduction of the *Notice of the General Office of the State Council on Establishing Leader Team of Government-Procurement-of-Service Reform* in 2016. The leading group is responsible for working on policies and measures for the reform, directing all regions and departments to advance the project, and addressing any difficulties between departments or across different fields.

The concept of "collaborative" governance

Born out of the necessity to move away from adversarial and traditional modes of governance (Ansell and Gash 2008; Rapp 2020), the practice of collaborative governance has considerably grown over the past two-and-a-half decades (Greenwood et al. 2021). As a result, a bewildering array of concepts, definitions, and modes of research emerged in the literature (Emerson et al. 2012). Although collaborative approaches to policymaking have been advocated in Western democracies in recent decades to close the growing gap between government and citizens (Batory and Svensson 2019), nowadays, the concept is applied in a wide array of institutional arrangements and governmental styles (Emerson et al. 2012), including China, and the heterogeneous and complex way in which scholars and practitioners use the concept greatly contributes to its fuzziness (Batory and Svensson 2019).

Ansell and Gash (2008), for instance, advance collaborative governance as an arrangement whereby public agencies engage with public and private stakeholders in a collective and consensus-oriented decision-making process (Kuhn 2016). According to Ansell and Gash, in general, collaborative governance follows several conditions: (1) public agencies start it; (2) participants include non-state actors that (3) engage in decision making and, thus, are not merely "consulted"; (4) actors meet collectively; (5) decisions are reached by consensus (even if not achieved); and (6) the focus of collaboration is on public policy or public management.

Against this fairly restrictive definition which emphasizes partnerships initiated by government parties, Emerson et al. (2012, 2) offer a broader definition of collaboration, acknowledging that collaborative governance should "carry out a public purpose that could not otherwise be accomplished". Although a consensus on the right definition of collaborative governance is missing in the literature, scholars tend to emphasize non-governmental actors taking part in the work of government and the potential of this partnership to address complex social, environmental, and urban planning issues, including sustainability challenges (Gray and Purdy 2018).

This lack of consensus on what makes up "collaboration" led scholars to refer to different things in a multitude of political contexts. As Batory and Svensson stress, the absence of any clear-cut definition opens the floor to different interpretations and, the concept thus risks turning into a buzzword or, worse, propaganda (see Batory and Svensson 2019). Policymakers can claim to be developing "collaborative" governance practices without genuinely involving the opinions of others (*ibid*) or worse, they can use the concept to advance authoritarian practices.

The arrival of a "collaborative" governance model in China?

For a long time, the Chinese state has limited the development of SGOs. Yet, as the state's traditional roles and responsibilities started being outsourced to non-state or private actors (Gao and Tyson 2017)—as shown above in the case of Shanghai—scholars increasingly emphasized the potential of "collaborative" governance (*hezuo zhili* 合作治理 in Chinese) in processes of policy making and policy implementation in China (Jing and Hu 2017; Ratigan and Teets 2019). I myself used the words "collaborative" and "cooperative" interchangeably in previous work analyzing waste management programs in Shanghai communities (Arantes et al. 2020). Although both words put the emphasis on actors working together, there is a difference between the two, and understanding this helps us capture the subtleties of China's current governance mechanisms.

Yijia Jing, as cited above, has been applying the concept of collaborative governance to the Chinese reality, with one of his many contributions is a volume that he edited, *The Road Towards Collaborative Governance* (2015b). Advancing several case studies such as environmental protection or disaster response, contributors to the book define the Chinese state's enthusiasm for engaging with non-state stakeholders as "collaborative" governance. According to Jing (2015b, 1), collaborative governance represents "the sharing of power and discretion within and across the public, non-profit, and private sectors for public purposes".

But is the practice of service contracting described above an example of so-called "collaborative" approach to governance? I previously stated that Shanghai was one of the first cities in China to undertake government procurement of social services on a large scale (Jing 2012; Jing and Gong 2012; J. Y. J. Hsu and Hasmath 2016). Ding Li 丁立, the vice-president of NPI, explained to me (Interview 24 June 2017) that the purchase of social services from SGOs by the government was partly developed by Lü Zhao, the founder of NPI. In 2007, Lü carried out a research project on government purchase of SGO service and provided detailed recommendations that were later accepted by the supervising government agency.

Since then, service contracting expanded to other provinces and, contrary to what might have been expected, created opportunities for local governments to take advantage of SGOs to meet the needs of groups with high service demands such as the elderly, the disabled, or the poor (Jing 2015b, 601). Still, the system has been favoring the growth of SGOs that performed well on issues that serve the interests of local governments (Y. Shen and Yu 2017, 13–14), in particular, since the Xi administration established new technologies of control and regulation (Charity and Overseas NGO Management Laws, for example).

There is no doubt, as Jing and Hu (2017) show, that such "collaborative" processes improve community governance and policy effectiveness because they increase citizen satisfaction while reducing government costs. But Jing also points to some limitations of such state-SGO "collaboration" practices. According to the Chinese scholar, a lack of competitive bidding and the CCP's emphasis on maintaining a stewardship model leads to collusion in the government-SGO collaborations (Jing 2015b, 10). As he stresses (2015b, 11):

"In the long run, Chinese leadership will increasingly feel the conflict between its demands for efficiency and control".

Other scholars, advancing similar thoughts, argue that this "Chinese" way of governance is more reflective of a "consultative" model rather than a "collaborative" one (Tsang 2015; Yan and Wei 2017; Ratigan and Teets 2019). Drawing on evidence from an affordable housing policy in Jiangsu, Ratigan and Teets (2019) show that despite an increase in "contracting out" encouragement and processes of community involvement in policymaking, "collaborative" governance was unsuccessful because the state agencies did not create meaningful roles for SGOs to engage in decision-making.

Recent evidence seems to suggest that the concept of Xi Jinping's "top-level design" (*dingceng sheji* 顶层设计) will continue to dominate, leading to a model of government-led policy design and service provision (Ratigan and Teets 2019). As Howell et al. (2021) have pointed out, during the Xi era, Hu-Wen's notion of "social management (*shehui guanli* 社会管理)"—which rationalized the changing roles of government and governed—evolved into "social governance (*shehui zhili* 社会治理)". The shift in terminology expresses Xi's determination to regain control over society. This signals, on the one hand, a deepening of governance through welfare, social work and social organizations or, as the authors put it, "a firm system of governance at a distance", and, on the other, an affirmation of power over SGOs (2021, 8).

This changing landscape can be seen during the period of Shanghai's first local experimentations (described above) up until to the current situation. Before the establishment of the Charity and the Overseas NGO Management laws, a grey area enabled local officials to contract work out to SGOs without full registration. The new laws granted rich cities like Shanghai, that enjoyed a considerable level of funding autonomy, a certain amount of leeway to use resources and side-step central level control. Let us consider one simple example: In Shanghai, the Municipal Bureau of Civil Affairs could make use of revenue generated from the social welfare lottery to finance a variety of social services (e.g., for children or the disabled). But the Municipal Bureau did not spend the money directly. Funds were first transferred to the District Bureau of Civil Affairs and then to the sub-district levels. At this point, it becomes difficult to trace, monitor, and control how, by whom, and for what the money was being used (Jing, 2012). Two of the observed organizations, for instance, claimed to be working with the local government despite not being legally registered (Turtle and Cat, see Appendix C).

The (re)centralization techniques, such as the Charity and the Overseas NGO Management laws in 2016, put an end to this kind of flexibility. Besides strict restrictions on funding and registration processes, it required local officials to form "blacklists" of groups that the government considers suspicious. Although local officials can still decide which groups to contract, the current political contexts leave little space for taking risks (Howell et al. 2021, 18). As is clear from this hostile environment, China is not heading towards a "collaborative" form of governance. The central government is readjusting and eliminating dissenting voices with a view to forming a unified one. Therefore, I contend that the current system is more representative of a "cooperative" model of governance. The difference between the two words lies in how parties reach their shared goals. Cooperation is when one group works together with another to support a singular goal—here, the attainment of an "ecological civilization". This cooperation, however, developed in an environment that only has very limited options available. The "green" consensus here assumes the role of main narrative not only for the shaping and control of environmental governance, but more generally for enhancing the CCP's hegemonic view.

Strengthening the role of SGO role in the post-political era

It is clear that since Xi Jinping came to office, the SGO landscape has transformed. Even though Hu's new institutions of public participation are still present, there has been a shift from fragmentation to a consolidation, resulting in a less room for contentious participation. Although the government accorded citizens and environmental SGOs the legal rights to sue polluters, new laws and regulations point to the government's willingness to adopt a more explicit divide-and-rule approach (H. Han 2018). Altogether, on the one hand, the government aims to incorporate any SGOs it considers useful and innocuous into an increasingly institutionalized system of "social governance"; on the other, it rejects and represses SGOs that are identified as threatening the party's authority. In the subsequent chapters, I also contend that the government's instrumentalization of environmental issues prevents environmental activists from resisting co-optation by the government machinery.

According to Slavoj Žižek, this kind of governance reflects a process of post-politicization, whereby politics or contention is increasingly replaced by hard and soft technologies of "policing" (administration) of environmental or other domains (Žižek cited in Swyngedouw 2019, 27). Those deemed irresponsible or "non-consensual" (because they contravene the hegemony of the CCP or do not align with developmentalism) are excluded and those able to sustain the top-down managerial and technocratic practices of sustainable development are co-opted by the regime. Since within this framework, SGOs cannot act contrary to the discursive imaginaries of "ecological civilization", their action perpetuates a common ideal or vision of environmentalism. Harvey Neo sees such governance practices as a "means to an end" (Jobin et al. 2021, 120) that ultimately strengthen political stability through non-confrontational means.

Focusing attention on SGOs' action at the local level thus offers new opportunities to assess the heterogeneous relations by which *environmental authoritarianism* is mobilized and enacted. It shows that SGOs are subject to the post-political reality of Shanghai and their action is reduced to the choreographies of promoting and cultivating a "green" community consciousness among the Chinese population. Such arrangements are proliferating within specific spaces at the sub-district-level (Chung 2018), like neighborhood communities, where SGOs act as tools in the state's strategy for "community-building (*shequ jianshe* 社区建设)" (Bray 2006).

The rationalization of sub-district governance

After its taking up of power in 1949, the CCP used workplaces as a basis to build a socialist society and provide social welfare services. Consequently, almost all social policies—surveillance, political education, or childcare—were administered through the workplaces, called "work units" (*danwei* in Chinese 单位). For example, the *danwei* system was key to the implementation of the one-child policy as citizens could be monitored through this work unit system.[12] Yet, at that time, a minority of the population lacked access to these services as they were not officially "affiliated" in the workplace system (Bray 2006), the basic spatial and social unit of urban China in the pre-reform era (Derleth and Koldyk 2004; Shieh and Friedmann 2008; Fulong Wu 2018). To remediate to this and as part of a push toward "community building", in the 1950s, neighborhood administrative organizations such as the Street Office (*jiedao banshi chu* 街道办事处) and residential committees (*jumin weiyuanhui* 居民委员会)—also called neighborhood or residents' committees—also became essential players in the party-state sub-district level structure (see Figure 3.3).

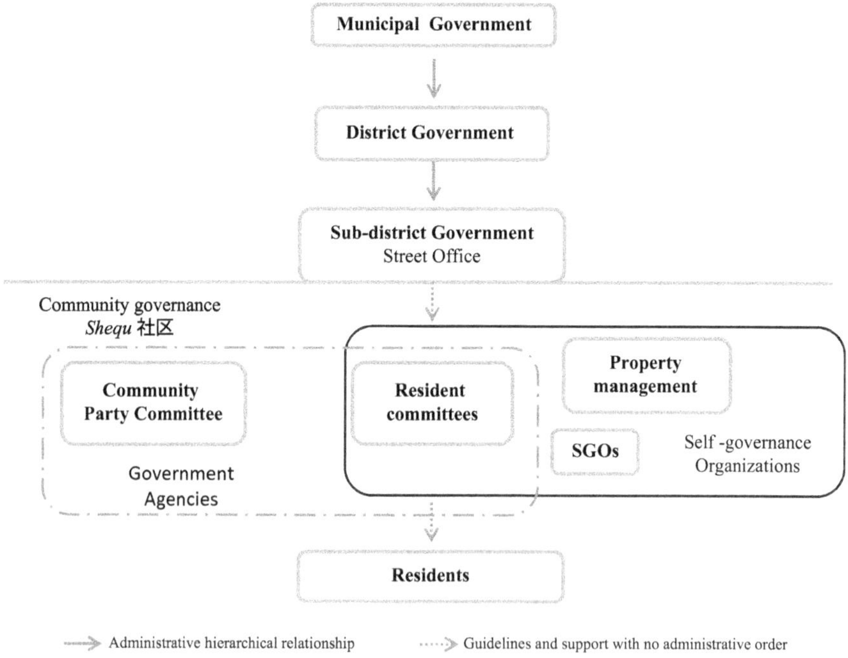

Figure 3.3. Shanghai's Municipal Government governance apparatus.

Since the 1980s, the Street Office and the Residents' Committees have been redesigned to replace the obsolete *danwei* system (as marked mechanisms replaced the previous planned economy) and respond to the state's aspirations to retake a central role inside communities (Heberer and Göbel 2011). It is mainly at this micro-local administrative level that SGOs develop their activities and that one can observe the political rationalities being set as part of the macro-background. As Trott (2016, 131) states (referring to China), the grassroots level is a microcosm of a country's problems.

As stated previously, to cope with the challenges of the acceleration of urbanization[13]—such as social and economic transformations, increasing numbers of protests, requests from civil society, or population diversification[14]—and the fiscal system of the early 1990s, Shanghai's municipal government contracted social organizations to provide welfare services. This is part of the state's effort to move the provision of social services from the state to the "community", or what Heberer and Göbel describe as the politics of "community-building", where discipline is enforced through self-discipline rather than coercive state organizations, such as the police (see 2011).

It is precisely to devolve some powers and functions to lower levels (to cope with increasing urban management demands) that Shanghai transitioned from a "two-tiered government, two-tiered management" to a "two-tiered government, three-tiered management" system (*liang ji zhengfu san ji guanli* 两级政府三级管理) (Chung 2018).[15] A "two-tiered government" refers to municipal governments and district governments, while "three-tiered management" added Street Offices to the government (Wan 2015; Fulong Wu 2018; L. Zhao 2019). Later, the system was extended to include Residents' Committees as agents of local government (Fulong Wu 2002). The development of Residents' Committees and Street Offices is said to be the driving force behind Shanghai's efforts at "community development" (Wong and Poon 2005; Wan 2015; Fulong Wu 2018). The initiatives discussed above (e.g., to remove barriers to SGO registration and operations) are part of this wider strategy.

Almost 20 years have passed since the Shanghai Municipal Committee launched the "two-level government, three-level management" system reform. Yet during the 2014 National People's Congress, Xi Jinping reinforced a desire to strengthen innovative governance solutions at the community level. As he stated: "The core is the people, the focus is on urban and rural communities, and the key is institutional innovation (*hexin shi ren, zhongxin zai chengxiang shequ, guanjian shi tizhi chuangxin* 核心是人、重心在城乡社区、关键是体制创新)".[16] As a response to Xi's call, in January 2015, the Shanghai Municipal Committee and municipal government jointly released the "1 +6" policy under its flagship project titled "Innovating Social Governance and Strengthening Grassroots Building (*chuangxin shehui zhili, jiaqiang jiceng jianshe* 上海创新社会治理、加强基层建设)". These initiatives put Shanghai at the forefront of "community-building" in China. Today, community-building has become a major part of China's governance system and a tool to advance consensual modes of policymaking around sustainable issues.

The *shequ*

In Shanghai, the pluralization of social services under the supervision of Street Offices was achievable through the establishment of a grassroots governance system. This is also the level where the governmentalities of social order and welfarism increasingly intersect and enable the state to permeate micro levels of practice, like the work of SGOs in communities. Micro levels have become the perfect site for *authoritarian environmentalism.*

> Neighborhood administrative organizations (e.g., Street Offices and Residents' Committees) are now utilized as engines for "social engineering" (Benewick et al., 2004; Yan & Guo, 2005). Based on their long-term presence in local neighborhoods as arms of the government, these pre-existing grassroots organizations have been easily transformed into alternative stations of service provision without threatening local state authority (Wong & Poon, 2005), largely due to the retrenchment of state-sponsored welfare provision. In addition to being the traditional agents of social control, grassroots organizations are now required to behave as both social service providers and the driving forces of community development (Choate, 1998; Wong & Poon, 2005). Street Offices and Residents' Committees have numerous responsibilities, including the creation of jobs for the unemployed, providing minimum support to vulnerable populations, taking care of urban sanitation and healthcare, and enforcing party policies that affect the daily lives of urban residents (e.g., family planning, social stability, and public-security-related works) (Saich, 2008).
>
> (Chung, 2018, 7)

Local officials made this management shift towards the "community" to ensure that service provision remained under the close watch of the Street Office (Chung 2018). The Street Office is the most frontline agency of the local government, and its core functions include "strengthening Party building, coordinating community development, organising public services, implementing comprehensive management, supervising professional management, mobilising social participation, guiding grassroots self-governance and safeguarding community safety" (L. Zhao 2019, 215). Street Offices need to respond to legislative constraints which either limit or empower their actions (X. Zhang et al. 2019, 93).

In Shanghai, grassroots governance is being developed with three primary objectives: party-building, community-building, and community management. As Zhao Litao (2019, 204) argues, the concept has evolved considerably because, in such a large megacity, the district is too large to act as the interface between government and residents.[17] To rectify the situation, decision-makers divided districts into several streets (*jiedao* 街道). The concept also explains the rising importance of the *shequ*—*she* meaning community or society and *qu* district or area. In the Memorandum on Promoting Urban Shequ Construction, *shequ* is defined as "a social collective formed by those residing within a defined geographic boundary", and the territory of a *shequ* is "the area under the jurisdiction of the enlarged residents' committee"' (Tang and Sun 2017).

The *shequ* is now designated as the basic unit of social, political and administrative organizations in urban areas (Q. Zhang et al. 2018), or the "organic cells of the city" (Rowe et al. 2016, 48). Even though the functions and responsibilities of *shequ* vary considerably across China (even within the same cityscapes) (Derleth and Koldyk 2004), they normally represent an urban precinct under the governance of the Street Office (Tang and Sun 2017), resembling a "type of city zoning" (Q. Zhang et al. 2018) (refer to images below). *Shequ* territories can be demarcated in three ways: the first is based on a single *danwei* compound, including residential areas; the second is based on a single, bounded residential compound or micro-district (*xiaoqu* 小区); and the third is based on a city block style, bounded by roads (Rowe et al. 2016).

Understanding these shifts at the *shequ* level is extremely important if we wish to understand the background of our case study in Chapter 4 and the context in which party-building is being developed in China's biggest cities. It is crucial to explore the central government's desire to "strengthen the grassroots" in order to apprehend why SGOs are developing the way they are. My fieldwork reveals that SGOs are more likely to grow within the delimited boundaries of the Residents' Committee. Chung (2018) argues that local states have been shifting the responsibilities of the provision of social services to lower-level authorities. Yet the budget for services still depends on the Street Offices which do not always have the resources to support all the necessary services. Thus, engaging with SGOs becomes a pragmatic strategy.

Figure 3.4. Shequ walls in Shanghai.

As our case study in Chapter 4 demonstrates, this is mostly because SGOs encourage volunteer participation, provide basic services, and enhance a shared identity among residents and a sense of belonging in the community. The case study also explains the importance of the promulgation of the Charity Law and the Overseas Management NGO Law for the central state. These laws responded to several issues that this self-governance model could entail if relying on a growing number of uncontrolled SGOs on the ground.

Naturally, when considering the growing number of non-state actors acting at a local level, the rapid redevelopment of urban management and the increasing number of citizens who live outside the scope of a centrally controlled workplace system, it became urgent for China's leadership to find new innovative ways to regain control over citizens' daily lives. Re-engaging with people at the *shequ* level without making them feel they are being controlled is part of that strategy (Chung 2018). An increasing body of scholarship is focusing on the party-state's strategies to consolidate its presence at this level (Derleth and Koldyk 2004; Shieh and Friedmann 2008; Almén 2018). While some have analyzed the growing presence of grassroots party secretaries on the non-public sector (H. Zhang 2018), others have focused on the government purchase of social services (J. Y. J. Hsu and Hasmath 2014; Yuen 2018), or the State's selective intervention in community disputes (Hu et al. 2018).

In some closely related work, other scholars have been focusing on the impact of a rising context of political participation at local levels. In response, some have referred to the Chinese system as a "fragmented authoritarianism" (Lieberthal 1992; Mertha 2009), a "consultative authoritarianism" (Teets 2013; Truex 2017), or a "deliberative authoritarian regime" (He and Warren 2011; Ma and Hsu 2018). Yet all these analyses tend to point in one direction: the current Chinese system has created space for more autonomy, bargaining, and participation. The studies also provide theoretical and empirical evidence of the way in which several governance mechanisms, typical of democratic regimes, are being implemented in authoritarian contexts. More importantly, they ask how these mechanisms help the party-state to stabilize and strengthen its authoritarian rule.

Although much has been said about the role played by local governments and a grassroots model of governance in terms of institutional transition, scholars until now have provided very little empirical proof on party-building efforts through SGOs development and rising "green" discourses in urban areas. In particular, research has not explored the way in which SGOs are being led to develop an array of political strategies and community-building alternatives that help the CCP to maintain power and legitimacy.

In the following chapter, I use the case study of ZeroWaste to further deconstruct how China's regime is taking advantage of such "discreet" and "obfuscated" feedback mechanisms at the sub-district level to reinforce a "green" consensus. I argue that these cooperative mechanisms with SGOs appear to be reinforcing Chinese citizens' "sense of belonging" and sense of responsibility in the face of growing social and environmental issues. This penetration of the party-state through indirect tools of convergence and control into the grassroots is made without making residents feel they are being controlled. Moreover, as further demonstrated in Chapter 5, convergence is increasingly difficult to contest. Still, these new mechanisms deserve attention as they embody the CCP's attempts to cope with the loss of legitimacy on environmental issues. This analysis also seeks to make a real contribution to an emerging scholarship on post-political forms of governance in Asia.

Conclusion

In this chapter, I aimed to highlight the role Shanghai has come to play as China's "green" face, focusing on how the city is reshaping its environmental governance model accordingly. In particular, this chapter focused on the favorable conditions developed by local leaders to implement a state corporatist model in Shanghai. Under this corporatist model, a variety of mechanisms, including cooperative governance arrangements with SGOs, are being developed with the aim of mobilizing social actors while keeping them under authorities' strict guidance. Such strategies can be observed at the subdistrict level through the work of SGOs, as I will show in detail in subsequent chapters. Rather than making use of SGOs and their differing views on how to cope with environmental issues, the "green" consensus stipulates a universal agreement on the interpretations of "good" environmental governance. Under such circumstances, the boundary between government and SGOs becomes permeable. It implies a common purpose, and joint action, or, what I describe as a cooperative framework of shared views, in order to achieve a singular (the state) objective.

Thus, the chapter pushes the need to reflect on the terms used to describe the CCP's new governance arrangements. The terms collaboration and cooperation, for instance, are often used interchangeably by scholars when describing state-SGOs relations in China. Yet I argue that a subtle difference in meaning between the two shows that China is more representative of a cooperative model of governance. Although Chinese leaders implemented several tools to enhance participation and increase public trust—such as the Environmental Impact Assessment Law (EIA) in 2002 which include experts and the public in the decision-making process (T. H. Y. Li et al. 2012) or enhancing the role of non-state actors as described above—the reality is that Chinese leaders work with other people to achieve their own goal as part of a common goal ("ecological civilization"). The role of SGOs, therefore, is directed towards responding to the state's objectives, and not the contrary. In fact, SGOs that are not in line with the state cannot register or get funding. Besides, relying on government for funding restricts the SGOs' capacity to negotiate. I develop this point in the next chapter by taking ZeroWaste as a case study. There we will see in particular how ZeroWaste's work in communities supports the goals of the municipal government.

The case studies presented in the next chapters will show how this cooperative governance actually functions as a "consensus" governance (Swyngedouw 2009, 604). Particular attention will be given to the role of SGOs in the colonization of the political arena by technocratic-managerial governance and how this renders alternative voices inaudible and, consequently, sustainability issues increasingly apolitical. Overall, this will show the proactive role taken by SGOs in making China's "green" consensus thrive.

Notes

1 As Weller states in his book *Discovering Nature* (2006, 59): "During the early decades after the Republican Revolution in 1911, cities—and Shanghai above all—had appeared as the height of all that was best about modernity. They had industry, intellectual ferment, and an utterly new style".

2 Ran suggests that the dysfunction of China's decentralized environmental governance structure is the outcome of a blame-shifting game between central and local governments. She states that "blame up" occurs when local officials blame officials at the center, while "blame down" is the other way around.

3 28 Jan, 2018 "Grossly Deceptive Plans: China's obsession with GDP targets threatens its economy", *The Economist*. https://www.economist.com/china/2016/01/28/grossly-deceptive-plans (accessed 27 November 2018).

4 In 2015, tens of thousands of people took to the streets of Shanghai to protest a proposal to relocate a paraxylene (PX) plant to the city's Jinshan District, with up to 50,000 people joining protests. This demonstration is said to be the largest protest so far in China against the production of PX, a compound used to manufacture polyester and plastic bottles. See Liu Qin (2015) Shanghai residents throng streets in "unprecedented" anti-PX protest. *China Dialogue*. https://www.chinadialogue.net/article/show/single/en/8009-Shanghai-residents-throng-streets-in-unprecedented-anti-PX-protest (accessed 14 August 2019).

5 See Xie's book *Environmental Activism in China* (2009) for more details.

6 Source: Shanghai Statistical Yearbook 2017.

7 See Zhang (2015) Government Procurement Law and Policy: China. *The Library of Congress*. https://www.loc.gov/law/help/govt-procurement-law/china.php (accessed 2 September 2019).

8 In a recent report, the World Bank set out a clear definition of what constitutes a competitive city. Briefly, it is a city that is growing fast, generating jobs, and increasing the level of income for its inhabitants. The report is available here: http://documents.worldbank.org/curated/en/816281518818814423/pdf/2019-WDR-Report.pdf (accessed 17 July 2019).

9 Interview with the Programme Manager of NPI, 21 December 2016, Pudong, Shanghai.

10 Interview with the Marketing and Communication Manager of NPI, 27 October 2016, Pudong, Shanghai.

11 For more information on the historical development of government procurement of public services refer to the report by Jia et al. titled "Final Report on Government Procurement of Public Services People's Republic of China", available at http://unpan1.un.org/intradoc/groups/public/documents/un-dpadm/unpan042435.pdf (accessed 23 August 2019).

12 Between 1980 and 2015, the Chinese authorities implemented a population planning initiative to curb the country's growth by restricting many families to a single child.

13 According to the National Bureau of Statistics, as of the end of 2018, the urbanization rate of China's permanent residents reached nearly 60 percent. This rapid urbanization transformed rural population into urban residents, but also included populations of migrants living and working in cities and towns, thus forming a transition between urban communities and rural communities and migrants from traditional rural areas.

14 There are several types of migrants in Shanghai, including people from the city centre who have migrated because of municipal reform, people from rural demolition and resettlement, or people coming from other provinces. See Shanghai shi chengxiang jiehe bu "zhuanxing shequ" de shequ zhili jiegou tan 上海市城乡结合部"转型社区"的社区治理结构探讨 [Discussion on the Community Governance Structure of "Transforming Communities" in the Urban-Rural Junction of Shanghai]. *Tongxie lunwen wang* 童鞋论文网. http://www.txlunwenw.com/xingzhenglunwen/2019031112598.html (accessed 10 September 2019).

15 Shanghai was the first city in China to adopt this model.

16 Ma Luping (2015) "Guanche luoshi xijinping zong shuji zhongyao jianghua jingshen 贯彻落实习近平总书记重要讲话精神 [Carry out the spirit of the important speech of General Secretary Xi Jinping]". State Council of the People's Republic of China, available at http://www.gov.cn/xinwen/2015-03/06/content_2828409.htm (accessed 12 September 2019).

17 Shanghai's population density has more than doubled since the late 1980s, increasing from 1,785 people per square kilometer in 1978 to 3,814 people per square kilometer in 2017. Refer to the Shanghai Municipal Statistics Bureau, available at http://www.stats-sh.gov.cn/tjnj/nje18.htm?d1=2018tjnje/E0201.htm (accessed 18 September 2019).

References

Almén, Oscar. 2018. "Participatory Innovations under Authoritarianism: Accountability and Responsiveness in Hangzhou's Social Assessment of Government Performance". *Journal of Contemporary China* 27 (110): 165–179.

Ansell, Chris, and Alison Gash. 2008. "Collaborative Governance in Theory and Practice". *Journal of Public Administration Research and Theory* 18 (4): 543–571.

Arantes, Virginie, Can Zou, and Yue Che. 2020. "Coping with Waste: A Government-NGO Collaborative Governance Approach in Shanghai". *Journal of Environmental Management* 259.

Batory, Agnes, and Sara Svensson. 2019. "The Fuzzy Concept of Collaborative Governance: A Systematic Review of the State of the Art". *Central European Journal of Public Policy* 13 (2): 28–39.

Bray, David. 2006. "Building 'Community': New Strategies of Governance in Urban China". *Economy and Society* 35 (4): 530–549.

Brehm, Stefan, and Jesper Svensson. 2017. "A Fragmented Environmental State? Analysing Spatial Compliance Patterns for the Case of Transparency Legislation in China". *Asia-Pacific Journal of Regional Science* 1 (2): 471–493.

Carrillo, Beatriz, Johanna Hood, and Paul IKadetz. 2017. *Handbook of Welfare in China*. Edward Elgar Publishing.

Chen, Sibo. 2017. "Environmental Disputes in China: A Case Study of Media Coverage of the 2012 Ningbo Anti-PX Protest". *Global Media and China* 2 (3–4): 303–316.

Chen, Yawei. 2007. *Shanghai Pudong: Urban Development in an Era of Global-Local Interaction*. IOS Press.

Chin, Yik Chan. 2016. *Television Regulation and Media Policy in China*. Routledge.

Chung, Yousun. 2018. "Continuity and Change in Chinese Grassroots Governance: Shanghai's Local Administrative System". *Issues and Studies* 54 (4): 1–27.

Derleth, James, and Daniel R.Koldyk. 2004. "The Shequ Experiment: Grassroots Political Reform in Urban China". *Journal of Contemporary China* 13 (41): 747–777.

Dong, Stella. 2001. *Shanghai: The Rise and Fall of a Decadent City.* HarperCollins.

Emerson, Kirk, Tina Nabatchi, and Stephen Balogh. 2012. "An Integrative Framework for Collaborative Governance". *Journal of Public Administration Research and Theory* 22 (1): 1–29.

Gao, Hong, and Adam Tyson. 2017. "Administrative Reform and the Transfer of Authority to Social Organizations in China". *The China Quarterly* 232: 1050–1069.

Gray, Barbara, and Jill Purdy. 2018. *Collaborating for Our Future: Multistakeholder Partnerships for Solving Complex Problems.* Oxford University Press.

Greenwood, Stephen, Laurel Singer, and Wendy Willis. 2021. *Collaborative Governance: Principles, Processes, and Practical Tools.* Routledge.

Han, Heejin. 2018. "Legal Governance of NGOs in China under Xi Jinping: Reinforcing Divide and Rule". *Asian Journal of Political Science* 26 (3): 390–409.

Han, Jun. 2016. "The Emergence of Social Corporatism in China: Nonprofit Organizations, Private Foundations, and the State". *China Review* 16 (2): 27–54.

He, Baogang, and Mark E. Warren. 2011. "Authoritarian Deliberation: The Deliberative Turn in Chinese Political Development". *Perspectives on Politics* 9 (2): 269–289.

Heberer, Thomas, and Christian Göbel. 2011. *The Politics of Community Building in Urban China.* Routledge.

Hou, Lulu, and Yungang Liu. 2017. "Life Circle Construction in China under the Idea of Collaborative Governance: A Comparative Study of Beijing, Shanghai and Guangzhou". *Geographical Review of Japan Series B* 90 (1): 2–16.

Howell, Jude, Regina Enjuto Martinez, and Yuanyuan Qu. 2021. "Technologies of Authoritarian Statecraft in Welfare Provision: Contracting Services to Social Organizations". *Development and Change* 52 (6): 1418–1444.

Hsu, Jennifer Y.J. J., and Reza Hasmath. 2014. "The Local Corporatist State and NGO Relations in China". *Journal of Contemporary China* 23 (87): 516–534.

Hsu, Jennifer Y J, and Reza Hasmath. 2016. "A Maturing Civil Society in China? The Role of Knowledge and Professionalization in the Development of NGOs". *China Information* 30, 10–16.

Hu, Jieren, Yue Tu, and Tong Wu. 2018. "Selective Intervention in Dispute Resolution: Local Government and Community Governance in China". *Journal of Contemporary China* 27 (111): 423–439.

Jia, Xijin, and Ming Su. 2009. "Final Report on Government Procurement of Public Services People's Republic of China". *Asian Development Bank.* www.adb.org/sites/default/files/project-document/64175/36656-01-prc-tacr-06.pdf

Jing, Yijia. 2010. "From Stewards to Agents? Competitive Contracting for Social Services in China". *Public Management Research Conference: Research Directions for a Globalised Public Management.* http://www.socsc.hku.hk/pmrc/papers/Session IV/Public Private Partnerships/Yijia JING _From Stewards to Agents.pdf.

Jing, Yijia. 2012. "From Stewards to Agents? Intergovernmental Management of Public-Nonprofit Partnerships in China". *Public Performance & Management Review* 36 (2): 230–252.

Jing, Yijia. 2015a. "Between Control and Empowerment: Governmental Strategies towards the Development of the Non-Profit Sector in China". *Asian Studies Review* 39 (4): 589–608.

Jing, Yijia. 2015b. *The Road to Collaborative Governance in China*. Palgrave Macmillan.

Jing, Yijia. 2018. "Dual Identity and Social Organizations' Participation in Contracting in Shanghai". *Journal of Contemporary China* 27 (110): 180–192.

Jing, Yijia, and Bin Chen. 2012. "Is Competitive Contracting Really Competitive? Exploring Government-Nonprofit Collaboration in China". *International Public Management Journal* 15 (4): 405–428.

Jing, Yijia, and Ting Gong. 2012. "Managed Social Innovation: The Case of Government-Sponsored Venture Philanthropy in Shanghai [政府管理下的社会创新：以上海市政府发起的公益创投为例]". *Australian Journal of Public Administration* 71 (2): 233–245.

Jing, Yijia, and Yefei Hu. 2017. "From Service Contracting To Collaborative Governance: Evolution of Government-Nonprofit Relations". *Public Administration and Development* 37 (3): 191–202.

Jobin, Paul, Ming-Sho Ho, and Michael Hsin-huang Hsiao. 2021. *Environmental Movements of the Anthropocene in East and Southeast Asia*. ISEAS Publishing.

Kerlin, Janelle A. 2017. *Shaping Social Enterprise: Understanding Institutional Context and Influence*. Emerald Group Publishing.

Kolhoff, Arend. 2016. *Capacity Development for Environmental Protection: Towards Better Performing Environmental Impact Assessment Systems in Low and Middle Income Countries*. Utrecht University.

Kuhn, Berthold. 2016. "Collaborative Governance for Sustainable Development in China". *Open Journal of Political Science* 6: 433–453.

Lee, Seungho. 2007. "Environmental Movements and Social Organizations in Shanghai". *China Information* 21 (2): 269–297.

Lei, Jie, and Kwan Chan Chak. 2018. *China's Social Welfare Revolution: Contracting Out Social Services*. Routledge.

Leung, Joe C B, and Yuebin Xu. 2015. *China's Social Welfare: The Third Turning Point*. John Wiley & Sons.

Li, Ding. 2016. "Public–Private Partnership in the Development of Social Entrepreneurship in Mainland China: The Case of NPI". In *Social Entrepreneurship in the Greater China Region: Policy and Cases*, 127–140. Routledge.

Li, Terry H.Y., S. Thomas Ng, and Martin Skitmore. 2012. "Public Participation in Infrastructure and Construction Projects in China: From an EIA-Based to a Whole-Cycle Process". *Habitat International* 36 (1): 47–56.

Lieberthal, Kenneth G. 1992. "Introduction: The 'Fragmented Authoritarianism' Model and Its Limitations". In *Bureaucracy, Politics, and Decision Making in Post-Mao China*, edited by Kenneth G. Lieberthal and David M. Lampton. University of California Press.

Liu, Huaxing. 2015. *Why Is Local Government Less Trusted than Central Government in China?* University of Birmingham. PhD Dissertation.

Ma, Deyong, and Szu-chien Hsu. 2018. "The Political Consequences of Deliberative Democracy and Electoral Democracy in China: An Empirical Comparative Analysis from Four Counties". *China Review* 18 (2): 1–32.

Mertha, Andrew. 2009. "'Fragmented Authoritarianism 2.0': Political Pluralization in the Chinese Policy Process". *The China Quarterly* 200: 995–1012.

Mok, Ka Ho, and Jiwei Qian. 2019. "A New Welfare Regime in the Making? Paternalistic Welfare Pragmatism in China". *Journal of European Social Policy* 29 (1): 100–114.

Ran, Ran. 2017. "Understanding Blame Politics in China's Decentralized System of Environmental Governance: Actors, Strategies and Context". *The China Quarterly* 231: 634–661.

Ran, Ran, and Yan Jian. 2021. "When Transparency Meets Accountability". *China Review* 21 (1): 7–36.

Rapp, Claire. 2020. "Hypothesis and Theory: Collaborative Governance, Natural Resource Management, and the Trust Environment". *Frontiers in Communication* 5: 28.

Ratigan, Kerry, and Jessica C. Teets. 2019. "The Unfulfilled Promise of Collaborative Governance: The Case of Low-Income Housing in Jiangsu". In *The Palgrave Handbook of Local Governance in Contemporary China*, 321–344. Palgrave Macmillan.

Rothery, Robert. 2003. "China's Legal Framework for Public Procurement". *Journal of Public Procurement* 3 (3): 370–388.

Rowe, Peter G, Ann Forsyth, and Har Ye Kan. 2016. *China's Urban Communities: Concepts, Contexts, and Well-Being*. Birkhäuser.

Salmenkari, Taru. 2017. *Civil Society in China and Taiwan*. Routledge.

Shen, Jianfa. 2018. *Urbanisation, Regional Development and Governance in China. Planning Perspectives*. Vol. 34. Routledge.

Shen, Yongdong, and Jianxing Yu. 2017. "Local Government and NGOs in China: Performance-Based Collaboration". *China: An International Journal* 15 (2): 177–191.

Sheng, Chunhong. 2019. *Guanxi and Local Green Development in China: The Role of Entrepreneurs and Local Leaders*. Routledge.

Shi, Si. 2017. "Research on Supplier Selection of Local Government Purchases in China: Case Study Of Shanghai". *IOSR Journal of Humanities and Social Science* 22 (06): 50–57.

Shieh, Leslie, and John Friedmann. 2008. "Restructuring Urban Governance: Community Construction in Contemporary China". *City* 12 (2): 183–195.

Smith, David Horton, and Ting Zhao. 2016. *Review and Assessment of China's Nonprofit Sector after Mao. Voluntaristics Review*. Vol. 1. Brill.

Swyngedouw, Erik. 2009. "The Antinomies of the Postpolitical City: In Search of a Democratic Politics of Environmental Production". *International Journal of Urban and Regional Research* 33 (3): 601–620.

Swyngedouw, Erik. 2019. *Promises of the Political: Insurgent Cities in a Post-Political Environment*. MIT Press.

Tang, Ning, and Fei Sun. 2017. "Shequ Construction and Service Development in Urban China: An Examination of the Shenzhen Model". *Community Development Journal* 52 (1): 10–20.

Teets, Jessica C. 2013. "Let Many Civil Societies Bloom: The Rise of Consultative Authoritarianism in China". *The China Quarterly* 213 (January): 19–38.

Teets, Jessica C., and Marta Jagusztyn. 2015. "The Evolution of a Collaborative Governance Model: Social Service Outsourcing to Civil Society Organizations in China". In *NGO Governance and Management in China*, 69–88. Routledge.

Trott, Stephen. 2016. "Grassroots Governance Reform in Urban China". In *Governance, Domestic Change, and Social Policy in China: 100 Years after the Xinhai Revolution*, 129–148. Palgrave Macmillan.

Truex, Rory. 2017. "Consultative Authoritarianism and Its Limits". *Comparative Political Studies* 50 (3): 329–361.

Tsang, Steve. 2015. "Contextualizing the China Dream: A Reinforced Consultative Leninist Approach to Government". In *China's Many Dreams*, 10–34. Palgrave Macmillan.

Tuan, Yang, Huang Haoming, and Andreas Fulda. 2015. "How Policy Entrepreneurs Convinced China's Government to Start Procuring Public Services from CSOs". In *Civil Society Contributions to Policy Innovation in the PR China*, 196–217. Springer.

Wan, Xiaoyuan. 2015. "Governmentalities in Everyday Practices: The Dynamic of Urban Neighbourhood Governance in China". *Urban Studies* 53 (11): 2330–2346.

Weller, Robert P. 2006. *Discovering Nature: Globalization and Environmental Culture in China and Taiwan*. Cambridge University Press.

Weng, Shihong. 2017. "Zhengfu xiang shehui zuzhi goumai gonggong fuwu de jianguan jizhi yanjiu 政府向社会组织购买公共服务的监管机制研究 [Empirical Study on Regulation Mechanism of Purchase of Service Contracting]". *Journal of Beijing University of Aeronautics and Astronautics Social Sciences Edition* 30 (4): 23–32.

Wong, Linda, and Bernard Poon. 2005. "From Serving Neighbors to Recontrolling Urban Society: The Transformation of China's Community Policy". *China Information* 19 (3): 413–442.

Wu, Fengshi, and Kin Man Chan. 2012. "Graduated Control and beyond: The Evolving Government-NGO Relations". *China Perspectives* 2012 (3): 9–18.

Wu, Fulong. 2002. "China's Changing Urban Governance in the Transition towards a More Market-Oriented Economy". *Urban Studies* 39 (7): 1071–1093.

Wu, Fulong. 2010. "Property Rights, Citizenship and the Making of the New Poor in Urban China". In *Marginalization in Urban China*, 72–89. Springer.

Wu, Fulong. 2018. "Housing Privatization and the Return of the State: Changing Governance in China". *Urban Geography* 39 (8): 1177–1194.

Wu, Jing, Yongheng Deng, Jun Huang, Randall Morck, and Bernard Yin Yeung. 2013. *Incentives and Outcomes: China's Environmental Policy*. SSRN.

Xie , Lei. 2009. *Environmental Activism in China*. Routledge.

Xie, Lei, and Peter Ho. 2008. "Urban Environmentalism and Activists' Networks in China: The Cases of Xiangfan and Shanghai". *Conservation and Society* 6 (2): 141–153.

Yan, Wang, and Chonghui Wei. 2017. "The Chinese Logic of Consultative Governance". *Social Sciences in China* 38 (3): 5–24.

Yang, Zhenjie. 2013. "'Fragmented Authoritarianism'—the Facilitator behind the Chinese Reform Miracle: A Case Study in Central China". *China Journal of Social Work* 6 (1): 4–13.

Yu, Jianxing, and Sujian Guo. 2019. *The Palgrave Handbook of Local Governance in Contemporary China*. Palgrave Macmillan.

Yu, Xiaomin. 2016. "L'entrepreneuriat Social Dans Le Secteur Non Lucratif En Chine. Le Cas de La Participation Innovante de La Société Civile Dans l'effort de Reconstruction Qui Suivit Le Tremblement de Terre". *Perspectives Chinoises* 3: 55–64.

Yuen, Samson. 2018. "Negotiating Service Activism in China: The Impact of NGOs' Institutional Embeddedness in the Local State". *Journal of Contemporary China* 27 (111): 1–17.

Zhang, Han. 2018. "Who Serves the Party on the Ground? Grassroots Party Workers for China's Non-Public Sector of the Economy". *Journal of Contemporary China* 27 (110): 244–260.

Zhang, Qi, Esther HiuKwan Yung, and Edwin HonWan Chan. 2018. "Towards Sustainable Neighborhoods: Challenges and Opportunities for Neighborhood Planning in Transitional Urban China". *Sustainability* 10 (2): 406.

Zhang, Xuefan, Jing Wang, and Li Xu. 2019. "Between Autonomy and Supervision: The Interpretation of Community Supervisory Committee Reform in Hangzhou, China". *Cities* 88: 91–99.

Zhao, Litao. 2019. "The Party in Grassroots Governance". In *The Chinese Communist Party in Action: Consolidating Party Rule*, edited by Yongnian Zheng and Lance L. P.Gore, 288. Routledge.

Zhao, Shuqing, Liangjun Da, Zhiyao Tang, Hejun Fang, Kun Song, and Jingyun Fang. 2006. "Ecological Consequences of Rapid Urban Expansion: Shanghai, China". *Frontiers in Ecology and the Environment* 4 (7): 341–346.

4 An iron fist in a velvet glove

Introduction

"What kind of trash are you? (*ni shi shenme laji* 你是什么垃圾)" became one of the most popular sentences in Shanghai after the new Domestic Waste Management Law came into effect on 1 July 2019. It was on everyone's lips, from retired citizens chatting in parks to social media influencers. Citizens greeted the program with many reactions, some applauding, others highly criticizing the complex and strict nature of the system which required Shanghai's 26 million residents, including commercial establishments, to sort their waste or risk incurring fines and perhaps even a lower social credit score.[1] As the first Chinese city to adopt legislation on domestic waste classification, Shanghai's "new era of compulsory waste-sorting" made the news in China and abroad.[2] It was somehow as if issuing the law was the first time that the government paid attention to waste management law. Yet, as this chapter explains, Shanghai's recycling system results from years of local experimentations between various stakeholders. Since 2012, for example, several environmental SGOs working on the waste front have run trials and explorations of effective ways to promote waste segregation in Shanghai's communities. Sponsored by the local government, they have developed tailored knowledge on waste management practices, from training, design, to operationalization.

As explored in the previous chapters, the need to cope with social and environmental problems following China's rapid economic development, led the Chinese government to gradually incorporate social groups into the arenas of social governance and the provision of public services (J. Wang and Wang 2018). To reach the city's "green" goals, everybody needs to be on board. These practices are not specific to China. Co-governance spaces have been established in a wide range of social and political contexts (e.g., Brazil, the United States or India) and policy areas (e.g., poverty reduction or school reform), and recognized as an effective means of improving accountability and governance (Ackerman 2004). A strong deployment of top-down management tools, consensus around environmental protection, and institutionalization of grassroots movements all have helped get waste management on track.

DOI: 10.4324/9781003231325-4

In this chapter, I focus on waste management as a case study to explore how the party-state is using SGOs to penetrate deeper into civil society. In particular, I emphasize the co-optation of SGOs as service providers and the instrumentalization of environmentalism as a mechanism to create novel forms of "green" citizenship. I use Actor-Network Theory (hereafter, ANT) to investigate the implementation of the "green account" (*luse zhanghu* 绿色账户), a municipal solid waste management cooperative program, through the mechanism of telling a story (Stanforth 2006). This analysis is developed against the backdrop of the strategies and techniques developed by Shanghai's municipal government to influence the daily lives of its citizens, here, more particularly, to make them recycle. The analysis will particularly highlight the role of ZeroWaste, an environmental SGO, and the different actors who work in cooperation, all with the common goal of recycling. This chapter will explore questions including: How is an array of different and conflicting political interests brought together to reinforce party-building at the grassroots? What is the role of SGOs? Are they helping the CCP to disseminate specific forms of knowledge? And can they reinforce the party-state influence at the grassroots? Although my reflections are based on the data gathered, the interpretations offered by people met in the field were an essential part of the analysis presented below.

Shanghai's waste "revolution"

China has become one of the world's biggest players when it comes to the production of municipal solid waste, especially in developed coastal cities such as Shanghai or Beijing.[3] Uncovering the tremendous amount of waste created by the country's rapid development was an "easy" challenge for Wang Jiuliang 王久良, China's famous photographer and filmmaker. Wang's documentaries, *Beijing Besieged by Waste* (2012) and *Plastic China* (2016), stunned viewers by showcasing the illegal waste dumps surrounding Beijing as well as depicting the hard lives of China's army of informal waste pickers. Wang's work received several awards and featured at many festivals around the world. In China, both films received a warm welcome but rapidly vanished from the web.

As Schulz (2019) put it, "something in *Plastic China* must have bothered Chinese censors". At first glance, the issues underlined in both documentaries look consistent with China's official discourse and growing greening practices. Wang's work makes a powerful statement against waste imports, in line with Xi's "hostile picture" of foreign forces. Indeed, I remember feeling particularly embarrassed about this during a brainstorming session that followed the screening of *Plastic China* organized by several Chinese activists in a small coffee shop in Jiading District in 2017. Although the movie had been censured, it was circulating among environmental activists who were sharing it by organizing small group screenings. As we were watching the movie, it was clear that much of the waste was coming from Europe. After the screening followed a round table. As my time to speak arrived, I could feel participants glaring at me, as if I were representing the voice of the "shamefaced" foreign exporters of waste.

Yet if we consider the movie in more detail, as Schulz (2019) argues, there are less obvious statements that could have touched a nerve with the Chinese censors. Because the documentary brings light to challenges such as poverty or social inequality, it discredits President Xi Jinping's statements on the "Chinese Dream (*zhongguo meng* 中国梦)" or "rejuvenation (*fuxing* 复兴)" of the nation. These were some of the concerns highlighted in the discussion. During the debate, a 12-year-old child who was attending the event with his parents said that he didn't understand why Yi-Jie, the 11-year-old girl actor who played the main character in the movie, was not attending school like himself (Figure 4.6). Although the discussion among adults had not focused on this specific point, the child's question reinforced Schulz's argument.

Shanghai produces enormous quantities of municipal solid waste,[4] about nine million tons per year to be precise.[5] A high rate of urbanization, fast technological advancements, and changes in the consumption behavior of its citizens mean this amount is rising at alarming rates (Zhou et al. 2019), as is the visibility of waste. To keep up with the amount of waste being produced, the government proposed burning it to produce energy. But incinerator projects have met high public resistance. China's citizens are aware of these projects' negative impact on health and have been actively protesting against construction in Shanghai, Zhejiang, Jiangsu, Hubei, Beijing, or Guangdong.[6] Moreover, despite a growing awareness among Chinese citizens,[7] consumption is not decreasing as of yet. Quite the opposite. As one of the fastest-developing cities in the world with a population capping 25 million, finding solutions is urgent for Shanghai.

Actually, the city has been trying to set up sorting and recycling plans for several years. The municipal government's initiative did not entirely serve their purpose. It was responding to central government directives aimed at setting strict standards at a national level. The first national legislation on solid waste management was issued in 1996 (Liu and Jin 2017). From the 2000s, eight pilot cities (including Beijing and Guangzhou) introduced waste campaigns, but many of these were complete failures. According to one informant (whom I will call here Mr. Liu, aged 67), even though measures were taken, people were not engaged with the process. For Mr. Liu, the project was deemed to fail for two reasons: first, because Chinese citizens were not aware of the consequences of rising quantities of waste; and second, because waste processors mixed all types of waste together after collection, so there was no point in recycling.

During my time in the field, I saw that local leaders had learned from these past mistakes, and their strategy had changed. The government could count on engaged activists eager to find solutions to the waste problem, and many grassroots initiatives emerged at the beginning of the 2010s, including one group called ZeroWaste (not their real name). The success of these groups in some communities caught the interest of local leaders. As explored below and in the following chapters, the problems surrounding waste management are far more complex than the mere act of recycling. The issue touches on environmental and urban management alongside less obvious matters such as migrants' rights to the city, inequality, economic growth, urban aesthetics, and other socio-environmental issues. This reality became evident during my experience in the field and during the many conversations I had with people working on the waste front.

Through observations and interviews with waste pickers, residents, and community leaders, I rapidly saw that waste could act as a good entry point to assessing politics and the ways in which governmental structures adapt to contemporary pressures. Besides revealing how Xi Jinping's "ecological civilization" (see Chapter 2) is conceived and implemented at a local level, I could observe the local state's logic on-site. While observing how local actors were implementing the "green account" program, what particularly sparked my interest was the way in which these "participatory" initiatives and discourses about a "beautiful city" were enhancing the party and the state's capacity for resilience.

The rest of this chapter focuses on assessing the political strategies behind Shanghai's "green account". This program, which would later become ZeroWaste, was established in hundreds of communities after a successful start as a pilot project run by a group of environmental activists. The chapter aims to understand the construction of China's grassroots, exploring not only how various interests came into alignment but also how an inclusive and cooperative governance model emerged in the process. I will, therefore, continue to delve into how China's leadership is trying to regain influence on the daily lives of its citizens through the increasing institutionalization of the role of SGOs in governance. Focused on primary empirical data, the next section will also discuss how such grassroots governance innovations are the result of complex processes of institutional and urban reconfiguration apparatuses.

The "story" presented below results from an ethnographic study conducted over several months. I followed the implementation of the "green account" in over 20 neighborhood communities (*shequ*) in an area between the core areas of Shanghai's city zone (Jing'an) to the north (Zhabei and Hongkou),[8] two zones where the project was being developed at that time. This long period of observation allowed me to observe how the aforementioned construction of grassroots governance is the outcome of negotiations among diverse actors as they attempt to extend their networks and maintain their complex relations through processes of *translation* in which all actors agree the network is worth building and defending.

To show how the construction of grassroots is being assembled, I develop an ANT-based analysis of the "green account" using Callon's four "moments of translation" (1986): (1) *problematization*; (2) *interessement*; (3) *enrollment*; and (4) *mobilization* (see Chapter 2). They will mark the key moments of our story. I will consider how dispersed actors mobilize, connect, juxtapose, and hold together in heterogeneous associations. This empirical consideration of the party-state's strategic development of sustainable communities is used to reveal how public participation is stimulated at the grassroots level.

Translating interests for a "clean" city

Environmentalism and civil society organizations (defined here as SGOs) are normally associated with social movements and democracy. The following analysis goes beyond such assumptions and questions the processes underpinning the development of environmental SGOs in authoritarian contexts. The analysis here draws on a case study of a cooperation between Shanghai's municipal government and over 30 environmental organizations. Some of the organizations do not fall into the category of SGO because they were created by the state, however my analysis focuses on ZeroWaste, an SGO which grew out of the grassroots movement where with whom I spent most of my fieldwork.[9] At the time, the "green account" was in its experimental phase but it became mandatory from July 2019. The program is to be rolled out in other big Chinese cities shortly.

The following section draws on ZeroWaste as a case study, considering from an ANT perspective what role cooperative governance (see Chapter 3) plays in the CCP's authoritarian resilience? Authoritarian resilience "in the making" is analyzed here from a micro-perspective, with a descriptive approach to resilience. Before delving into the analysis, I will briefly present the case study.

The "green account" program

To respond to the various problems caused by waste in the Metropole, the municipal government of Shanghai was looking for a suitable program that could, on the one hand, make citizens take part in recycling practices while, on the other, regain people's confidence by showing that their concerns are being heard. In addition, the mega-city also needed to keep up with a growing number of environmental SGOs such as ZeroWaste, that specialize in the waste sphere. As will become clear, to develop its waste segregation policy, the Shanghai authorities engaged and intertwined a diverse array of political interests. The previous chapters presented several key points explaining the Chinese leader's long-term strategy in dealing with the emergence of new actors in the environmental sphere. I defined this strategy as *environmental authoritarianism*, which builds on a dual-use of top-down enforcement and bottom-up "participatory" mechanisms, with a strong narrative surrounding the need to "jointly green the country".

As we will see, the "green account" reflects this cooperative strategy. Shanghai's municipal waste sorting plan focuses primarily on household participation. To encourage the public to partake in waste segregation at source, an incentive model was developed to push citizens to engage in recycling practices. This is, of course, a complex thing to do because social contradictions and conflicts are increasingly diversified and normalized (H. Zhang et al. 2019). Social governance relies on a multiplicity of subjects and, thus, becomes more "elastic" and "flexible".[10] For these reasons, it is a challenge for Chinese leaders to cope with an increasing array of different actors, each defending their own interest.

The "green account" program is based on the promotion of dry and wet waste classification. Before the program's implementation, each community-building floor has a single dustbin where all types of waste were collected. There was no segregation at source. Yet this didn't mean that recycling was non-existent. An army of informal waste collectors, mostly migrants, earn a living in the city by collecting valuable waste. An example here is Mrs. Wei—a migrant from Guangxi who collected waste every day in the building where I lived for several months. She explained that plastic bottles, valued at 2 RMB per kilo, or copper, valued at 20 to 35 RMB per kilo, as well as aluminum, valued at 6 to 7 RMB per kilo, are particularly "profitable". As a result, substantial amounts of waste were in fact being recycled. Other types of waste were mixed and directed to waste processing centers.

The "green account" of ZeroWaste completely redesigned this "informal" system by rewarding good, civic-minded behavior. According to the Director of ZeroWaste (Interview 26 April 2016), the aim of the campaign was not to charge fees but to generate incentive mechanisms that would encourage citizens to become "green" citizens. The "green account" card fueled incentives of this type as residents scored points every time they segregated their waste, and they could later redeem these points for prizes.

Problematization

A *problematization* means to "formulate a problem". The way in which a problem is framed is essential to catch the interest of a diversity of actors and make them move from a singular position—where each party follows their own interests—towards a cooperative one (Callon 1986). Put differently, the promoters of the idea seek to find actors who may have a similar interest and convince them that this suggestion is the best option. In doing so, those leading the project make themselves indispensable to the process. By formulating the problem and the actors involved, we implicitly delineate who is concerned and why.

When the convergence of interests emanating from heterogeneous entities is successful, it means that the actors accept the *problematization* and they become *enrolled*. Yet there are as many *problematizations* as there are actors. So, to assure the perpetuity of a project, *interessement* mechanisms must preserve the interest of actors in the long run. The *problematization* of the "green account" is to ensure waste segregation at source. Yet a *problematization* "can be a tricky business, because resistance or deviance from certain actors can subvert an entire project or bring it to naught" (Schouten et al. 2014). Finding good *interessement* strategies is thus crucial.

The government's wish to introduce fresh forms of waste management was already highlighted above. To become indispensable in this process, the municipal government needed to ensure that its definition of dealing with waste was recognizable to others. This characterizes what ANT defines as an *obligatory passage point*. The waste management program under consideration here is what Bruno Latour termed a heterogeneous assemblage.[11] It consists of human actors such as consumers, trash collectors, property management, neighborhood community, governmental officials, and so on. Nonhuman actors include collection schedules, regulations, propaganda posters, recycling bins, food, and odors from decomposition processes, along with others.

The cooperative function of the "green account" is the *obligatory passage point*. It links and translates the interest of different actors, such as residents, environmental organizations, trash collectors, and local entities. This *obligatory passage point* is the starting point to which a series of secondary points attach. It gives form to a perfect context for the municipal government to develop its goals. As Shin (2016) argues: "An OPP (obligatory passage point) can be thought of as the narrow end of a funnel, that forces the actors to converge on a certain topic, purpose or question".

Secondary *obligatory passage points* help us understand the importance of including certain players in the program. For instance, to make citizens recycle, it is mandatory to find incentives—here, the creation of a reward system was indispensable. Moreover, to regain the confidence of citizens and mitigate for previous negative experiences, the proponents of the project need to make sure waste is not mixed after collection. Additionally, the program should be accessible. In previous attempts at recycling, the government gave the task of segregating to the communities but provided little information on how to do so. A lack of resources, knowledge, and guidance led the communities to focus on superficial goals, merely motivated by a desire to avoid government inspections. Providing the appropriate resources is another *obligatory passage point*. So, to ensure that the program is followed, local actors need to guide, train, and supervise the residents. Ensuring that waste is correctly segregated is essential for keeping the residents interested in the long run.

From this short list of *obligatory passage points*, one can already identify key actors, as visualized in Figure 4.1. Here we see the *problematization*, the main and secondary *obligatory passage points*, as well as the actors linked to those passage points. The devices or strategies that need to be mobilized to create a cooperative waste management program also become clearer. These *obligatory passage points* are not exhaustive, however, as many *secondary passage points* could entail third passage points, and so on. The idea here is rather to present how a *problematization* generates an *obligatory passage point*.

Interessement/Enrollment

The *problematization* defines the principal actor and how that actor connects other indispensable actants to negotiate a common focus. The *interessement*

Figure 4.1. Main and Secondary Obligatory Passage Points of the "green account".

phase begins when the main actor tries to assign a specific role to other actors and stabilize it—that is to say, to ensure that it is impossible for them to return to the patterns or situation from the past. This phase consists of a series of processes taken by the leading actor to convince others to be "locked into place" (Callon 1986). In the present case, the *interessement* phase started around 2011 when the Shanghai Municipality pushed its districts to reduce their domestic waste by five percent annually.[12] The classification of municipal solid waste involves multiple levels of government: The Shanghai Greening City and Bureau (municipal level), each district Landscaping and City Appearance Bureau, the Street Office and the resident/neighborhood committee. It follows the state hierarchy described in Figure 3.3 (Chapter 3).

Our analysis here focuses on the Street Office or community level (*shequ*). Even so, it is important to note that activities developed at the micro-level are also designed to respond to agendas set at higher levels (Li et al. 2019). Xi Jinping himself alleged that waste segregation represents a "new fashion" (*xin shishang* 新时尚), and a "social civilisation" (*shehui wenming* 社会文明) indicator.[13] Focusing our gaze on the *shequ* is a way of observing the direct consequences of Xi's top-level design, in particular considering: how are *shequ* framed to respond to China's *environmental authoritarian* model. Exploring this level reveals important insights to better understand state-society interactions. Limiting our analysis to key players such as the state is not enough to make sense of a socio-technical network, nor to impose a solid and sustainable innovation, even in an authoritarian context.

Zero Waste

In this analysis, ZeroWaste represents the leading *intermediary* of the network. An *intermediary* "transports meaning or force without transformation" (Latour, 2005, 39). For Latour, the actions of *intermediaries* are predictable but can become *mediators* by transforming, translating, or distorting the initial message over time, and vice versa, *mediators* can become *intermediaries* (*ibid*). ZeroWaste is "locked in", since its official registration in 2012 as a civil non-enterprise unit (*minban fei qiye danwei* 民办非企业单位 in Chinese, see Chapter 2). Before creating ZeroWaste, its founders worked in another organization they had started in 2000 with other Fudan University law graduates. That organization focused on a multiplicity of social good causes, but, starting from 2009, the Director of ZeroWaste, in collaboration with her colleagues, gradually focused on promoting community waste segregation. They incentivized citizens to trade in waste that was harmful to the environment in exchange for gifts and prizes. In 2011, the team came into the media spotlight after their program got outstanding levels of participation.[14] In the space of only three months, about 90 percent of residents in the community where they were based sorted 11 different categories of waste, even though the city government only required them to separate into five categories.

According to ZeroWaste's Director (Interview April 2016), at some point, the need for more professionalization led the group to "spontaneously" establish (*zifa chenglile* 自发成立了) ZeroWaste. And yet, despite successful outcomes, legal and administrative hurdles put them under stress. Before the abolition of the dual management system, registering as a non-profit was difficult because organizations needed to be affiliated with a government department (see Chapter 2) and hold 100,000 RMB in funding.

In 2012, the head of a District government Department of Science and Technology supported their registration, and after collecting the required 100,000 RMB, they successfully registered as an NGO. For the present analysis, the registration can be seen as a powerful *interessement* strategy. Moreover, the relevant departments of the Shanghai Municipal Government elevated the ZeroWaste model to one of the three models of waste classification to be conducted in the city. In effect, ZeroWaste was *enrolled.* At that time, Huang, another of my informants in his thirties, had been working for ZeroWaste for a year. Speaking to me during a volunteer activity in a community, Huang explained that they received around 50 percent of their funding from the government. Much of this funding was being used to replicate their recycling model in Shanghainese communities. Since the organization was created, they established waste management projects in Shanghai's different districts, running trials and exploring effective ways to promote waste management and waste reduction. By generating new solutions to a social problem that surpassed the service capacities of the government, their activities ballooned all over the city. Their recycling program went from one community in 2011 to over 229 in 2016 (Interview 26 April 2016), and as a result, by the end of 2017, 137,548 households were *enrolled* in the program.

Residential Committee

The residential committee implements central and municipal directives at the local level. Yet, as a director of a neighborhood residential committee in

Figure 4.2. Entrance of a Residential Committee.

Jing'an District explained (Interview 26 March 2017), they receive limited material and financial support. Led by Street Offices and functioning at the lowest level of Shanghai's municipal government, the committees are essential to the functioning of China's political system (Bing 2012) because they impose decisions made at higher levels to citizens (X. Zhang et al. 2019).

Even more challenging than a lack of subsidies is a lack of enthusiasm coming from residents, argued ZeroWaste's Director. I previously highlighted how prior recycling attempts were a failure. Adding to these pressures, residential committees rely on the willingness of each property management team to handle the waste infrastructure. Yet, the interests of property management and those of the neighborhood committees do not always match. It is then the residential committee that acts as a negotiator between residents' complaints and property management. By bringing support to the residents' committee and responding to several of their problems (e.g., lack of engagement, fund, or conflicts with the property management), the residents' committees were easily *enrolled* in the initiative, to return to our ANT analysis.

Property management

The property management handles everything that is related to waste facilities. They need to respond to the city's appearance and environmental sanitation administration, from hiring qualified people to treat waste, building the recycling amenities to developing a good community environment. This means that the new obligations on waste management are a challenge for them. To keep up with waste regulations, property managers need to implement quick and effective changes, and their role is extremely important. A ZeroWaste worker stated that it is difficult to develop a successful recycling program when property management is not "on board". *Interessement* mechanisms needed to be developed to *enroll* these actors.

By joining the "green account" and developing a successful recycling program, a *shequ* can more easily receive a "civilized community (*wenming xiaoqu* 文明小区) model quarter" award (see Figure 4.3). To be recognized as a "civilized community" by the Shanghai municipal government, the *shequ* needs to meet several criteria.[15] For instance, vehicles need to be parked in an orderly way, and streets must be clean, while green spaces or tree coverage are also a plus. Of all the requirements set by the municipal government, the *shequ* needs to get at least 90 points out of 100. Having a good recycling program is one of the requirements for receiving the award. Earning this recognition means that the community is an exemplar "civilized" neighborhood. In the long run, winning the award equates with rising property prices and rents. Of course, this affects residents not able to afford the escalating prices of these houses which, according to Pow (2009, 129), reinforces the moral superiority of the inhabitants of such "civilized quarters" as compared to others.

Volunteers

Volunteers work to ensure that residents sort their waste. The ones I met were, in most cases, older citizens. ZeroWaste's workers argued that volunteers engage in the program because it makes them feel valued. It gives them a

Figure 4.3. Entrance of a "model quarter" Neighbourhood Community.

sense of accomplishment. They don't receive any reward except respect from other residents, yet they are increasingly important in sustaining the Party's grassroots governance system.

Working closely with the residents of their communities, the volunteers engage in all kinds of activities, from organizing karaoke sessions, teaching classes, offering help, to making sure everyone stays home during COVID-19. Volunteers are, thus, good *intermediaries,* especially for supporting the government will and interests. As far as the "green account" was concerned, volunteers received training from ZeroWaste. According to ZeroWaste's Director, they act as a sizeable network and resource pool that the party-state can depend on.[16] It is important to note that there are volunteers (over-50s, in general) from diverse backgrounds and some are closely attached to the ranks of the Communist Party.

Even though my analysis does not explore the party-state's power on/through volunteers, this group embodies an important role in the government's "community-building" strategy mentioned above (see Chapter 3). For instance, I witnessed volunteers escort the trash collector during collection rounds to check whether residents were following the procedures. Acting as a sort of watchdog, the role of the volunteers resembles that of Maoist China's work-unit people (crucial to implementing the one-child policy). Because the "green account" reinforces the volunteers' role inside their communities, ZeroWaste easily *enrolled* them despite the job being quite tough. Besides accompanying the trash collector, volunteers take shifts to check residents at the recycling points, and they also scan the residents "green account" cards.

Trash collector

Usually, each residential community has a trash collector (*huanwei gong* 环卫工, in Chinese). He/she deals with everything related to waste management. Being a trash collector is demanding and requires long hours of hard work. Waste needs to be managed seven days a week. The trash collectors are usually migrants who specialize in recycling and reuse by selling waste to recycling plants, factories, and reprocessing centers. One of the trash collectors I talked to, Mr. Li, was already living in a residential community for several years. He lived with his wife in a "small house" (*xiao ceng* 小层) arranged by the property management, and he saw this new recycling scheme as an advantage because it improved his living conditions as working hours are shorter.

Mr Li was quite satisfied with the "green account" program. Before the implementation of the "green account" he had to collect trash in every building, sometimes floor by floor. With around 20 to 30 buildings in a community, each comprising several floors, it was hard work. Nowadays, he makes just two rounds each day. People come to meet him and dispose of their wet and dry waste in his truck. Sometimes volunteers join him and help scan the "green account" cards. Mr. Li described how the recycling program improved his working conditions considerably, for example, offering better equipment and sanitation supplies.

Non-human intermediaries

To solidify the actor-network, the waste sorting plan depends on other, non-human elements. According to Lourenco and Tomael (2018), technical artefacts (here, e.g., the points on the card, the new sanitary locations to wash hands) also influence human actions and relationships. These tools can also be important to help us understand how new ideas become accepted and new mechanisms and protocols are adopted and integrated by a group. They act as an important apparatus to keep the interests of all the actors aligned via negotiations or by weakening other actors who might turn against the goals of the network. These non-human elements also operate as *intermediaries.* Below, I identify the most relevant, but the list could be longer. Presenting them enables us to undercover how the regime is mutually using social and material elements to produce knowledge and or certain actions.

- The "green card": In 2011, the Shanghai Municipal Solid Waste Classification and Reduction Promotion Joint Conference Office and the Bank of China jointly created a scan code classification card. The card enables residents to score points with the end goal of receiving gifts and prizes. The points can be redeemed for milk, soup, toothpaste, phone cards, tickets for tourist spots, or to pay utility bills.
- Information tools: Different strategies are used to target diverse generations. While traditional methods such as community blackboards, notices in public or in the local newspaper target an older generation, according to ZeroWaste, younger generations are more responsive to approaches using electronic and social media. WeChat is a strong *intermediate* in the network. It is used to publicizes and share information, goals, or objectives, but also accomplishments. Several of the organizations I observed used WeChat daily to share an ideal vision of the organization.

Figure 4.4. Community trash containers after the implementation of the "green account".

The collection points: Strategic collection points in the community replace the single floor bins normally used in traditional communities. To create a good and appealing environment, ZeroWaste instructed property managers to install water facilities so people can wash their hands. ZeroWaste's workers argued that cultivating a friendly and warm environment makes people feel their lifestyles are being improved, and this is essential for changing residents' behavior and habits. In some communities, big character banners underlined government discourses, such as: "participate in waste-sorting, create a beautiful community (*canyu lese fenlei gong chuang meili shequ* 参与了色分类共创美丽社区)". The "green account" program was also integrated within the government's propaganda tools. This explains, in part, why several of the residents thought ZeroWaste was a governmental organization—I will return to this point later.

Mobilization

The section above demonstrates how it is through a diversity of actants that our socio-technical network expands and solidifies itself. Yet this does not mean that, sooner or later, *enrolled* actants may not refuse to play the "role" appointed to them. For instance, ZeroWaste workers deviated from the official discourse when arguing that the government doesn't put enough effort into regulating environmental issues. Also, in some of the observed communities, the residents didn't engage with the program, but continued consistently mixing all wastes in the recycling points. Of course, informants might not tell me this during interviews, but spending time in the communities enabled me to observe acts of resistance. Although most of the *shequ* I visited were respecting the "green account", some were not.

When I asked ZeroWaste's director why the program was not successful in all communities, she argued this usually occurs in communities where one of the key actors didn't follow their role, in particular, when residents' committee leaders are not popular or don't fully engage with the program. A strong sense of community was also depicted as crucial by ZeroWaste workers I talked to during training events. Chatting with residents, I became aware that communities with transient populations were less likely to attain the high recycling levels of communities where "native" Shanghainese prevailed. Migrants (non-Shanghainese) were, on several occasions, accused by some of the volunteers of not making enough effort to integrate themselves in the communities. Although this goes well beyond the objectives of this book, the fact they did not speak Shanghainese was debilitating to migrants in some of the observed communities. The aforementioned ZeroWaste worker, Huang (originally from Hunan), resigned from his post after a year and a half. He said he could no longer bear being discriminated against for being an "outsider". Broadly speaking, therefore, various factors could affect the successful implementation of the program. Yet, according to ZeroWaste's director, the "green account" turned out to be effective in most of the communities where it was established. The municipal government could *translate* (create alignment in interests) the interest of key actors. Here, the local government, ZeroWaste, the property management, the waste collector, the volunteers, and the neighborhood committee.

But the socio-technical network is not fixed and new actants can emerge through time. Following the continuity of the network by tracing its evolution and its actants can reveal important details about authoritarian regimes' micro-dynamics. In some communities, surveillance cameras will soon replace volunteer shifts. New actants are (continuously) being introduced to make sure others keep acting according to the *problematization* which helps to deal with potential betrayals. The Charity Law, established in 2016, also puts a strain on ZeroWaste and other SGOs because they need to fight over (scarce) resources. ZeroWaste has to compete with other SGOs to get access to governmental funds and the competition among them is high. Frequently, people working in SGOs alleged that their program was better than others, while critics pointed to the fact that other groups might maintain strong ties with the government (or acting government puppets). There was a conflicting opinion from organizations who saw themselves as genuine SGOs versus other fake organizations that acted as arms of the state (normally called government non-governmental organizations GONGO); this, even though in practice, all groups received governmental funding.

ZeroWaste's guiding role in environmental governance

> Studies that locate the driving force either in the state or in the civil society miss out on a basic fact that the Chinese Communist Party (CCP) plays a central role in steering and coordinating the efforts in community building and grassroots governance.
>
> (Zhao 2019, 199)

Zhou et al. (2019) claimed that the participation of SGOs in social governance innovation is promising. It solves a variety of problems by building consensus and linking daily life with public issues: (1) developing local expertise and appease conflict; (2) helping to optimize the governance structure; and (3) ultimately, engaging public participation by making residents feel that their quality of life is improving. Zhou et al. (2019) argue these three key aspects help the local government to engage more efficiently in "community-building" practices, enhancing the core function of the *shequ*.

But these "community-building" activities—which I see as reinforcing citizens' engagement in participation mechanisms while strengthening party-building—were being mixed with recreational activities such as karaoke or playing mah-jong. In the field, I observed leisure to be an increasingly important propaganda tool. I took the image below during an activity organized by ZeroWaste. Residents came to redeem their points while volunteers were performing a karaoke session. I witnessed similar events on various

Figure 4.5. Karaoke session organised in a Shanghainese community during a "Green Account" activity.

occasions in several other communities. I would argue that these activities have many uses for local government and party officials because they function as a favorable space for the development of "self-governance" practices. These activities not only respond to an urgent environmental issue, but they also boost social harmony. They were also highly visible, thus reinforcing forms of "disguised" state authority, and new regulations and the bureaucratization of community-based control are thus incubated through such techniques.

Zhou et al. (2019) highlight similar research findings. With growing social problems and a push from the Chinese government to embrace cross-sector collaboration, some SGOs identify niches where a strong need for service is needed (Ewoh and Rollins 2011). Especially for environmental SGOs, a period of increasing environmental awareness (e.g., Expo 2010) coupled with a favorable institutional environment, opened the way to new collaboration opportunities. ZeroWaste developed in one such favorable environment. Its innovative way of dealing with waste attracted the government's attention. The importance of SGOs like ZeroWaste has been recognized by the government itself: "When it comes to guiding residents to take part in waste classification, government agencies and social forces are working hard, and the role of social welfare organizations in this cannot be ignored".[17] Since the beginning of the 2000s, mostly in developed coastal areas, local governments used the political rationale of promoting service contracting in urban spaces to meet rising service demands.

Yet local government interest in cooperating with ZeroWaste changes the way residents perceive the organization in the political sphere. Although a group's visibility and role in tackling environmental challenges is boosted by government support, SGOs may end up working as buffers between the state and citizens. In this sense, as Rutland and Aylett (2008) assert: "the translation of interests, while it allows diverse goals to be furthered to some extent, always involves a kind of betrayal of enrolled actants". During this process, interests are gradually translated into broad-based support for specific local actions (Rutland and Aylett 2008). SGOs act as bridges of social communication between the government and Chinese society because they promote mutual understanding, "open" and "transparent" decision-making processes, and "scientific" decision-making quality. The "green account" serves as a striking example of these mechanisms, but the same could apply to other initiatives such as urban gardens projects. SGOs increasingly act as *intermediaries* to set good citizen behavior. Their work shapes a desirable and socially responsible middle-class by imposing on each citizen standards of acceptable behaviors vindicated by their "expert" knowledge.

The "green account" serves as a perfect municipal waste management plan with the aim of changing city-dwellers' behavior without "dictatorial" measures. Shanghai is an interesting case in point here, because its residents seem more resistant to enforced mechanisms (at least, at the time of fieldwork). Although it is not an absolute distinction, in Shanghai, for example, citizens tend to ignore the mandatory metro security checks, whereas in other cities, such as Beijing or Shenzhen, these are not voluntary and Shanghai's "walk-around security" practices would not be tolerated. During fieldwork, I was also told by several actors (citizens, scholars, university students) that Shanghai is more cosmopolitan, dynamic, and, thus, more liberal and detached from the central government. The recent protest in the city after Beijing dropped the phrase "freedom of thought" from the charter of Shanghai's Fudan University is another example of dissent.[18] Although such non-compliance will probably diminish with the use of facial recognition and the social credit system, it is an interesting particularity of the city.

The "green account" shows that Shanghai's government is changing its governing mode through softer and "discreet" control mechanisms in society. Their cooperation with SGOs is key to this strategy. The case study explored here shows a "state-guided model" for the development of SGOs. I explained in previous chapters how various regulations, norms, and platforms ensure that SGOs develop in the Party's image. The case of ZeroWaste shows that these novel approaches and mechanisms impact the relationships between different actors to the state's advantage. During fieldwork, I myself witnessed how organizations were sometimes frustrated by not being able to develop the program as they would like to. Indeed, to win state funds, any proposal needs to closely match the state's requirements.

Following Shanghai's law on waste sorting, Geoffrey Chun-fung Chen, a lecturer in the department of China Studies at Xi'an Jiaotong-Liverpool University in Suzhou, described Shanghai's clean-up act as "authoritarian environmentalism". According to Chen, "It's not environmentally based consciousness from the bottom. It's a sort of eco-dictatorship, a very strange but somehow effective mode of governance".[19] Although few scholars explore SGOs' new roles in this strategy, as I try to prove in this chapter, this is a valuable means of exploring how authoritarian institutional transition mechanisms are being applied at a local level.

Are SGOs playing with the devil?

The empirical data gathered in this book show how SGOs open new perspectives regarding the concept of citizen participation[20] in China. They shed light on channels of information and socialization which reduce the government–society gap. The activities developed by environmental SGOs provide opportunities for citizens (with related interests) to interact and develop new opinions. They create social experiences between unrelated individuals, ranging from chatting online, doing volunteer work, or starting cooperative projects.

In addition to my observations with ZeroWaste, I took part in many other activities organized by other grassroots organizations working on environmental issues. Ranging from beach clean-ups, environmental classes in primary schools, "do it yourself" workshops or documentary nights, the activities revolved around the fact that, as individuals, our actions can have a big social impact. I participated, for instance, in collective activities organized by BlueOcean, an organization which focused on protecting the marine environment. In May 2016, over 60 people took part in the clean-up beach day. BlueOcean held the activity in a small fishing village about one-and–a-half hour's drive from Shanghai city center. The activity focused on collecting waste and sharing pictures on social media, such as an image of a dead bird's stomach filled with plastic. The group placed particular emphasis on including participants in future decision-making processes. After each activity, participants were asked about their feelings: What do you think can be improved? What did you enjoy the most? Or, as an individual, what can you change to positively impact society?

The activities encouraged specific forms of public advocacy and citizen participation (or citizenship). Because citizens' actions take on additional significance, social agency is reinforced. In a society where individualistic morality has become the norm, SGOs awaken collective values. By actively taking part in the organization's activities, a sense of community takes form, as claimed by three of my interviewees who worked at Chinese environmental SGOs. China has little tradition of civil society; citizens are used to receiving administrative commands and therefore expect the government to take responsibility for everything. Here, I argue that SGOs change this frame by accentuating individuals' environmental identities.

Another example of SGOs' role in awakening collective values is the organization of informal screening of censored films, as mentioned above. On such occasions, environmental activists from different organizations come together to reflect on current environmental issues with volunteers or interested members of the public. After the movie, organizers invite each participant to share his/her thoughts. It is important to highlight that this type of activity is extremely informal and based on personal relations. Still, this shows that SGOs can create more open and accessible environmental governance mechanisms, because they know how to monopolize information and how to

Figure 4.6. A child shares his thoughts after the screening of Wang Jiuliang's movie, *Plastic China*, organised by three environmental SGOs.

use social media. This does not mean, however, that their actions are not being "colonized" by state discourses. During my fieldwork, my WeChat account was filled with daily news/events/activities/shares of information from the organizations I was following. Almost no day went by without an event, a conference, a grassroots project gathering, or a film screening. SGOs use such events to find volunteers, to share their actions and achievements, and expand their network. People can participate and share ideas about current and future projects or give suggestions aimed at improving the organization. These events represent a space where everyone feels that their opinion counts. More importantly, the action of SGOs helps promote the idea that individual action makes a difference. My claim is that these actions enhances Chinese citizens "self-governance" conscience.

Their activities attracted well-off urban classes in particular, and this point will be further developed in Chapter 6. Apart from older citizens who mainly act at a *shequ* level, and who may come from a variety of backgrounds, the younger volunteers came from (generally) a middle- or upper-middle-class background and either had children (and wanted to transmit "good" values to them, or were concerned about their future in the face of growing pollution and climate change), or were looking for life-fulfilling experiences. Some were Chinese citizens working in foreign companies. My informant, Mr. Wu, is representative of such volunteers. For example, he worked for an international company that allocated him four hours per month to engage in social impact activities. He chose to volunteer with BlueOcean because his family comes from a fishing village. Typically, volunteers such as Mr. Wu have more time, more skills, and a particular motivation to engage (either personal or professional). At first glance, therefore, SGOs function as platforms or "intermediate organizations" for people (with common objectives and concerns) to come together and realize their social freedom (Woldring 1998, referring to

Tocqueville's free association concept). Yet do their actions enable forms of grassroots collective action in a Tocquevillian spirit?

I argue here that cooperative projects or the fact that SGOs funding depends almost exclusively on government resources in fact increases "friction". In her book *Friction: An Ethnography of Global Connection*, Anna Tsing (2011) uses this term as a metaphor to describe diverse and conflicting contemporary social interactions. Tsing takes as a fact that many environmental movements can expand their work without challenging the apparatus of global corporatism (which is responsible for the problem) by developing a non-frictional acceptance of that system's rules. This is exactly what I observed in the field.

I argue that two major factors cause this. First, SGOs tend to focus on professionalizing their work in one key area, be that waste management, plastic in the sea, urban gardens, or education. As Barker (2010, 10) put it, they become "specialised knowledge professions". Drawing on their expertise, SGOs collaborate with citizens and receive their input and feedback which increases the organization's efficiency. Yet their know-how is easily "co-opted" by the state machinery because the more the organization develops expert knowledge, the more this knowledge becomes apolitical. As a result, it becomes difficult to differentiate SGOs from the state apparatus. When I asked my informant Huang about this issue (WeChat conversation, March 2017), before they resigned from ZeroWaste, they responded that:

> In Shanghai, organizations working in waste sorting issues depend on the government to purchase services for funding. Therefore, it is normal that residents think ZeroWaste is part of the government. Another reason is that ZeroWaste is now undertaking government projects, so it is understandable.[21]

Given these circumstances, I found that even people closely working with ZeroWaste don't know how to define their organization. Below, I present a conversation in a WeChat group between a volunteer and three staff from ZeroWaste. The original conversation dates from May 2017; it was in Chinese and here is translated by me—all names have been changed.

VOLUNTEER: The Street Office invited the Linyi community to lead a propaganda film on community party-building. I explained to them we are organizing waste management activities under the guidance of ZeroWaste. They asked what is the nature of ZeroWaste? ☺

ZEROWASTE WORKER 1: We are a grassroots civil society (*minjian caogen* 民间草根).

ZEROWASTE WORKER 2: Social organization (*shehui zuzhi* 社会组织), a private non-enterprise unit (*minban fei qiye danwei* 民办非企业单位).

VOLUNTEER: A public welfare social organization (*gongyi xing shehui zuzhi* 公益性社会组织公益). The government purchases the services of ZeroWaste to lead the waste classification and other environmental activities in the community. ☺ ☺ ☺

ZEROWASTE WORKER 2: Secretary Li said it well.

The results of a question survey conducted with Chinese colleagues in 20 neighborhood communities in Shanghai (Arantes et al. 2020) corroborate this conclusion: by keeping a low profile, SGOs don't generate high social recognition among the population. In the study, most respondents taught the activities developed by SGOs were part of the state apparatus. Their action ends up being "colonized" (Barker, 2011) by the state. The same happens to the residents' committees.

Similar to *shequ* institutions, the activities of SGOs cultivate active and responsible citizens. As Wan (2015, 15) argued: "To mobilise them (retired citizens), the *Shequ* institutions embrace a hybrid of Maoist and Confucian discourses to strengthen their values as senior citizens". To encourage residents to live in disciplined ways, the discourses emphasize individual "activeness" or "progressiveness". *Shequ* institutions use volunteers and draw on their deep social bonds inside the neighborhood community to maintain social order, whereas SGOs use the "individual" rhetoric to push citizens to act in a certain way. Figure 4.7 visualizes this distinction. SGOs use discourse strategies similar to those of the state, which transfer the environmental "burden" to individuals.

For instance, in ZeroWaste's official website, a Mao-era propaganda poster accompanies the following message: "The waste segregation of each community can be pushed forward, you are not alone, with already five years of experience in communities, ZeroWaste is with you". A Maoist propaganda imaginary accompanies the message with the aim of reinforcing the value of individual actions. The picture in Figure 4.7 on page 100 is a WeChat post published by a ZeroWaste worker stating that "To become a good citizen, recycling is an obligation". These statements influence collective goals and particular identities and reinforce the CCP's rhetoric of "ecological civilization".

Identifying actors and following their actions helps expose the dynamics of institutional change and how it creates resilience through adaptation. By deconstructing the "green account" socio-technical network, the ways in which authoritarian resilience takes root within individuals and organizations is reinforced along with the institutional and technological infrastructure surrounding them. Still, ANT treats all actors as symmetrical, which is problematic in an authoritarian regime such as China. To remediate this shortcoming, in the final section of this chapter I employ Foucault's concept of *governmentality* to propose the idea that the state is extending specific forms of knowledge to encourage specific behaviors among the population.

Figure 4.7. ZeroWaste staff WeChat post demanding residents act as "good citizens".

Governance beyond the State

The empirical evidence outlined above shows that, to achieve its objectives, the state relies on incorporating new actors in the arena of governing. By developing new cooperative mechanisms with SGOs at the local level, for instance, individualized responsibility is activated. This leads to a rearrangement of the relationship between the state and civil society. Foucault's work on government and *governmentality* offers a reflexive space to further elaborate on the processes of "vertical decentralization" towards localized forms of governance and, thus, mitigates the limitations of ANT for rendering unequal relations of power. As stated by Rutland and Aylett (2008, 631), *governmentality* "looks at how a given piece of music comes to be performed by multiple actors but tells us little about how the piece was chosen or composed". I have show, above, using Callon's four moments of *translation*, how the music is chosen and composed. But how should we read this changing role of the state and reorganization of government techniques? What changes does it bring to the power relations between the state and society?

To respond to these questions—renewed relations between the state and civil society actors—we need to delve deeper into the various practices and strategies aimed at structuring and shaping the possible action of participation. This approach will help to identify the "price tag" of the involvement of new actors in the governing arena (Lemke, 2001, 202). As several studies have shown around the globe (Kent 2009; Thörn and Svenberg 2016), a variety of actors (e.g., SGOs) can hold individuals liable. Governments, businesses, and environmental organizations do promote varying techniques to support the idea of voluntary agreements and self-regulation for dealing with rising environmental issues.

As Foucault's concept of *governmentality* attests, power relations are characterized as "conduct of conducts". Scholars of *governmentality* have shown that a transfer of operations of government to non-state actors do not signal a decline of state sovereignty *per se* but an extension of government from formal to informal techniques of government (Lemke 2015, 84–85). As such, studies of *governmentality* have been helpful in illuminating the "soft" mechanisms of power and how individuals can be governed through "controlled autonomy". The case of China is important because it shows the authoritarian and normalizing underpinnings of participation under *environmental authoritarianism.*

Governmentality as "environmentality"

The concept of *environmentality*, or *environmental governmentality*, emerged in the 1990s when several theorists such as Luke, Darier, and Rutherford, appropriated a Foucauldian analysis to explore understandings on environmental issues, considering how human subjectivities are enacted regarding the environment. Their scholarship is of great importance for this book because it seeks to understand how *governmentality* is being used to structure how questions of environmental management—from dealing with the climate urgency to our individual habits of recycling—are being performed by those in power. As Rutherford put it, this has become a useful concept to think through the ways the environment is not only a biophysical reality, but a site of power (Rutherford 2017).

The notion of *governmentality*, then, is useful in its environmental articulation because it addresses political questions about the "conduct of conduct" by individuals and groups in their interactions with society, their environment and themselves (Luke 2016). Eco-governmentality defines "a modern political rationality which tends to 'conduct the conduct' of human beings by planning their surroundings" (Taylan 2017, 262). Here, I analyze this construction both in terms of the creation of an object of knowledge and a sphere within which certain types of intervention and management are created and deployed to enhance the government's broader aim of managing the lives of its citizens. So, again, this concept frees my analysis of previous work on *authoritarian environmentalism*, aiming to focus on how authoritarian mechanisms are used to settled environmental practices.

An interesting aspect of the notion of *governmentality* is that it gives new insights for the analysis of authoritarian contexts, which are normally viewed as a single issue. In contrast to viewing China as a monolithic political system where the Communist Party projects its will upon the population, *governmentality* takes into consideration the decentralized and ad hoc nature of power which "acts from *within* and is enforced by the subjects themselves" (Rutland and Aylett 2008). In other words, governance is not constructed through rigid structures of sovereign power but distributed across a multitude of self-governing actors (volunteers, SGOs, residents' committees) (Rutland and Aylett 2008, 630).

As I showed through the ANT analysis in the previous section, China is developing innovative ways of governing its population by acting upon its living environment. Its ability to reach a diversity of actors through the development of various networks and the use of specific "technologies", as explored by Hoffman (2009, 119), is understood as an important tool for the CCP to stay in power.

> An examination of the rationalities shaping green and sustainable city-building in present-day China illustrates an important shift away from the privileging of state planning and central government direction and towards less direct techniques of governing, multiple sources of authority and a diversity of governmental actors. Contemporary Chinese governmentality encompasses not only state actions, but also market logics of value, transnational discourses of environmentalism and global forms of statistic-based knowledge. Recognizing the extension of techniques beyond the "Party-state apparatus" to include market and community players, what Gary Sigley (2006a: 503) refers to as the distinction between government (zhengfu) and governance (zhili), helps us to understand how the environment is being targeted as a domain of action in China today.
>
> (Hoffman 2009, 119)

Here, I content that concepts such as *governmentality* are key to analyzing the government's changing practices at the *shequ* level and the modus operandi of *environmental authoritarianism*. Based on empirical data gathered in Beijing, Wan (2015) demonstrates that self-governing projects are organized between residents and the members of Residents' Committees to pass down political orders and mobilize bottom-up participation through senior citizens. These studies center their attention on the CCP's efforts to regain a certain level of control over a "banner of community-building" by the means of liberal governmental strategies (Wong and Poon 2005; Gorman 2016). In this light, scholars have given great attention to population policies (Greenhalgh and Winckler 2005), prostitution controls (Jeffreys 2006), *shequ* institutions (Wan 2015), online activism (Gorman 2016), or the social credit system.

Little scholarship has focused on environmentalism as a domain of governing. Apart from Lisa Hoffman's (2009) study, which argued that China's contemporary urban sustainability includes both "practices of government" and "practices of the self", much remains to be understood about the state's methods of managing its control over the environmental sphere. To analyze more deeply the integration of environmentalism and sustainability into China's rationality of governing, the "green account" program acts as a useful lens. As previously shown, local government's urban sustainability practices "aim not only to construct 'an environmentally friendly society' (*huanjing youhao xing shehui* 环境友好型社会)" (Xia, 2006) but also to cultivate "'well-dressed young people [who] mingle' in public squares and understand conservation (Li and Xue, 1999: 32)" (Hoffman 2009, 108).

To achieve their goal of "ecological civilization", the state requires the compliance of a range of different actors including municipal and district planners, government officials, waste management specialists, non-state actors, or volunteers. To capture the new dynamics in China's transiting grassroots society (Wan 2015, 16), the notion of *governmentality* is important because it helps to analyze how environmental governance is practiced from within. For Rutland and Aylett (2008), the key step in processes of governmentalization happens when the local state achieves its objectives by disseminating certain forms of knowledge to produce self-governing citizens. This approach is epitomized, for example, by the introduction of a new waste classification system which calls on residents to change their behavior.

Put simply, the institutional changes and top-down restrictions arrangements frame the behavior of SGOs. After being "confined" to a specific role, the state apparatus uses SGOs to reinforce the development of responsible citizens. ZeroWaste's action is directed towards creating self-reflexive individuals who take responsibility for their own waste. Following the initial participatory and volunteer phase in July 2019, segregating waste became mandatory. Yet, the discourse changed. The way in which waste issues are presented is constructed around everyone's (ir)responsibility. The implicit objectives are reducing and recycling, and achieving these are made possible through everyone's efforts. As M. Wang (2019) argues, waste is important because the very act of throwing away waste carelessly reflects a decline of a sense of responsibility among the Chinese population caused by authoritarian forms of urban governance which did not promote active participation in the public space.

Thus, although the CCP places itself at the forefront of developing an "ecological civilization", the reality is that the burden of dealing with the climate is increasingly redirected onto citizens' shoulders. Meanwhile, the party-state links strong leadership and environmental protection to justify assertive forms of governance. Key to the bridging of stricter regulations and individual conduct is the post-political condition, which determines what is considered legitimate and of collective interest. As this chapter has tried to show, waste management does not shatter the consensus advanced by the authoritarian state. The same is true for causes such as protecting the ocean or the creation of urban gardens. In other words, cooperative governance requires that SGOs work on an issue that is acceptable and desirable. All demands sidestepping the "green" consensus as planned by Chinese leadership, are not to be tolerated. In this case, ZeroWaste actions become a means to an end to: (1) change citizens' behavior; (2) enhance China's commitment in environmental performance; and (3) cut through an informal recycling system which is outside state control.

The structure of Foucault's analysis is interesting because its theoretical framework allows us to explore how the micro-politics of authoritarianism impact citizens' everyday lives (Dunlop 2019), and helps expose how the structures of consensus take shape through particular technologies of governing (Swyngedouw 2019, 11), such as cooperative governance.

This double strategy of "conducting the conduct" of citizens and creating grids of environmental histories—translating the priorities of the state into the goals of groups (e.g., SGO) and individuals—is key to the creation of a "green" consensus. These processes give voice and power to certain groups but continue to exclude those who are not in line with the mainstream discourse. As I will show in Chapter 6, such governance strategies have several limitations, yet, we can still observe that China's leaders have made a strategical choice: giving (some) people sufficient space to express their grievances and keep the illusion of freedom, but never enough to allow people to grow too confident. Understanding this *translation* will help China's observers better assess how the CCP is adapting to its severe environmental crisis. Nevertheless, Shanghai has several characteristics that differentiate it from other big cities in China and more studies are needed to assess whether such practices are being replicated in less developed areas.

Conclusion

This chapter presents a detailed picture of the development model of Shanghai's municipal government "cooperative" approach which was described in Chapter 3. I explain how this model is leading SGOs to be increasingly co-opted by the state machinery. Taking the "green account" as a case study, I argue that these mechanisms are used to develop self-governing citizens. The "green account" shows that giving new roles to non-state actors does not equal space for democratization. Although at first glance, it seems that the development "cage" of SGOs has expanded, the reality is that the bars of that cage are reinforced, their role will become trapped within technologies of "expert" administration and management (Swyngedouw 2019, 8).

By developing in a controlled space, SGOs are being instrumentalized and used as tools to restructure state-society relations. They are particularly helpful in creating congruence between the interests of the state and those of the individual. To develop this argument, I draw on my ANT analysis. This methodology is a useful tool to capture and make sense of the new state–society relations which influence the party-state resilience capacity (Heeks 2013). By "following the actors", I highlight that by taking into account a broad constellation of actors and successful processes of *translation* and associations between them, one can observe how authoritarian resilience emerges in a particular place, time, and scale. In this light, I showed how an SGO can stabilize certain ordering processes such as waste segregation. While SGOs act as important new state-society intermediaries to *black box*—treat as "knowledge which is accepted and used regularly as a matter of fact" (Yonay 1994)—certain governing practices, it is also important to pay attention to the heterogeneous networks of actants (humans and non-humans) that interact and engage in negotiations, discussions, consensus building, actions, and counteractions to either establish or block the implementation of governing innovations (Guzman and De Souza 2018). This is particularly important at the *shequ* level.

Exploring micro-local dynamics uncovers important evidence about the actual practice of government (what or who is governed, and how), including those with no formal involvement in the political system. Empirical evidence from our case study suggests that the role of SGOs in the *shequ* institutions is increasingly central to implementing government policies and projects, but also to new claims about the role of citizens. As such, citizens become key instruments for the local government to maintain order. We can thus conclude that the growing number of government-SGOs partnerships are effective in promoting governmentalizing actions (Luke 2016), and also in supporting the growing presence of the party at the grassroots level (party-building).

While ANT enables us to deconstruct local governance dynamics and analyze how innovation is constructed, it gives little information on how power relations shape the behaviors of others. Foucault's concept of *governmentality* is useful here for two main reasons. First, because it helps demonstrate how certain environmental problems promote, rather than contradict, the official government solutions to environmental problems. Second, it provides a better perspective to look at the power of the government's everyday practices, which are not given enough attention. As demonstrated empirically here, diffuse forms of power and different governmentalities are applied between and within state-level government agents, municipal authorities, local waste workers, and local communities to implement and (re)shape governance. The analysis points to the complexity of urban environmental governance and everyday politics in which action repertoires (e.g., environmental and pollution-related topics) are underpinned by the use of environmental discourses and, the role of citizens, to pass down political orders and activate "ecological" behaviors. These strategies fit very well with Xi Jinping's "ecological civilization" model, which encourages all citizens to actively participate in environmental protection.

Several studies have applied the concepts of *governmentality* and ANT to analyze how local environmental governance is developed (Rutland and Aylett 2008; Blok 2014). In this chapter, I have brought these approaches together to describe how, to maintain legitimacy, the CCP's leadership mobilizes a multitude of actants and discourses into various hybrids or assemblages to stabilize a certain mode of governing—in this case, as related to environmental issues. These approaches help highlight the various ways in which non-state actors are involved in policy development and implementation in authoritarian settings. As indicated by this case study, when actors embrace a "green" development model, it becomes difficult for conflicting or opposing discourses to argue against the state. It also becomes extremely difficult for SGOs engaged in environmental issues not to take advantage of this "green turn" to gain access to resources. Yet being co-opted by the state machinery is not without any consequences because doing so reinforces Xi Jinping's discourses on creating a "beautiful China" while diminishing a group's advocacy agency.

Yet neither the community nor the state act as a reified and unitary whole. Each of the state layers and local actors have their interests, and these should not be ignored in our understanding of power, interests, and governance in authoritarian regimes, especially when looking at a society as complex as China. Communities' intra-dynamics are as diverse and important as those of the state. SGOs act as important allies because they are successful at developing an agency at a local level. Therefore, even though their actions are increasingly instrumentalized by the state apparatus, the power of SGOs should not be completely ignored. Also, while the chapter has focused on showing how SGOs are co-opted by the state, it is important to stress that SGOs' actions should not only be interpreted as going against the state. Some see following the state's agenda as a legitimate means of attaining their goals.

As one leader of an SGO stressed during a weekend spent picking waste outdoors in the mountains, in Hangzhou: "We are in a time when many windows were closed, but we should strive and continue as they could open when you less expect it". Even though the case study in the following chapter strongly indicates that SGOs are becoming arms of the state, one should not disregard SGOs' capacity for resilience. As a prelude to the coming chapter, it is important to understand the "pessimistic" assumptions presented here with some nuance. Despite the restrictions faced by SGOs, they are constantly developing spaces to increase their independence, for instance, by choosing to develop market strategies. Yet, as we shall see next, embracing the market, while creating new development opportunities, does not help them fully escape the trap of post-politics.

Notes

1 See Wu Yixiu (2019) "Shanghai's compulsory waste sorting begins". *China Dialogue*. https://chinadialogue.net/en/cities/11349-shanghai-s-compulsory-waste-sorting-begins/ (accessed 17 August 2019).

2 For instance, the *South China Morning Post* article titled "Shanghai begins new waste sorting era, as China eyes cleaner image", or *The Guardian* article titled "'A sort of eco-dictatorship': Shanghai grapples with strict new recycling laws".

3 Source: Organisation for Economic Co-operation and Development statistics on Municipal waste, Generation and Treatment, available at https://stats.oecd.org/Index.aspx?DataSetCode=MUNW (accessed 3 January 2022).

4 Municipal solid waste, more commonly called rubbish, consists of everyday items discarded by the public.

5 See Ministry of Environment and Ecology (2018) "2018 Nian quanguo da, zhong chengshi guti feiwu wuran huanjing fangzhi nianbao 2018 年全国大、中城市固体废物污染环境防治年报 [Annual Report on Environmental Pollution Prevention and Control of Solid Waste in Large and Medium Cities in 2018]", available at http://gts.mee.gov.cn/gtfwgl/gtfwjkglgg/201901/P020190102329655586300.pdf (accessed 6 October 2019).

6 Zhang Pinghui (July 2019) "Thousands protest in central China over waste incineration plant". *South China Morning Post*. https://www.scmp.com/news/china/society/article/3017386/thousands-protest-central-china-over-waste-incineration-plant; or Michael Standaert (April 2017) "As China Pushes Waste-to-Energy Incinerators, Protests Are Mounting". *YaleEnvironment 360*. https://e360.yale.edu/features/as-china-pushes-waste-to-energy-incinerators-protests-are-mounting (accessed 7 October 2019).

7 Riots and discontentment increased from the mid-2010s. Chai Jing's documentary *Under the Dome* (2015) played a role in this increase. After the release of the movie, everything related to incinerators received far more attention. Incinerators release high quantities of PM 2.5, a particle that provoked considerable controversy and even forced China's authorities to adopt new methods for disclosing environmental information (A. L. Wang 2017).

8 Housing reforms in China have popularized a new type of settlement that is often constructed in an enclosed building form. The residential compounds either form an area that is visually segregated from the rest of the city but not completely shut off, or it is completely enclosed by big walls and closed to outsiders. Normally, these building forms have around 10,000 to 15,000 residents.

9 See Zhang Jun (June 2019) "Zhong qing bao guanzhu shanghai pojie lese weicheng: Guanjian kao qingniáan canyu 中青报关注上海破解垃圾围城：关键靠青年参与 [The China Youth Daily pays attention to breaking the garbage siege in Shanghai: the key is to rely on youth participation]". *The paper.* https://www.thepaper.cn newsDetail_forward_3610100 (accessed 12 October 2019).

10 Original passage translated by the author: "Shehui maodun ji chongtu ye riqu duoyang hua, changtai hua, shehui shenghuo de bu queding xing riyi zengqiang, zhe yiweizhe shehui zhili chuangxin biran yao jiezhu duoyuan zhuti de liliang, qi fangshi lujing ye ying gengjia juyou tanxing he linghuo xing 社会矛盾及冲突也日趋多样化、常态化，社会生活的不确定性日益增强，这意味着社会治理创新必然要借助多元主体的力量，其方式路径也应更加具有弹性和灵活性".

11 Bruno Latour defines hybrids as "mixtures […] of nature and culture" (Latour 1993, 10). They are tangled beings, assemblages of different entities that cannot be divided into two poles (Latour 2004, 24). Looking at hybrids allows new visibility of "matters of concern" (*ibid.*) which include non-humans, humans, as well as the producers of assemblages (Zimmer 2010).

12 Shanghai Municipal People's Government (2016) "Lese fenlei de 'Shanghai mochi'—lüse zhanghuyinfa quango guanzhu 垃圾分类的'上海模式'—'绿色账户'引发全国关注 [The 'Shanghai model' of waste classification—'green account' raises national concern].", available at http://www.shanghai.gov.cn/nw2/nw2314/nw2315/nw5827/u21aw1097675.html (accessed 14 October 2019).

13 Zong shuji guanxin de baixing shenbian shi | lese fenlei: 'Xin shishang' di meili zheshe 总书记关心的百姓身边事 | 垃圾分类：'新时尚'的美丽折射 [The general secretary cares about | Waste classification: the beautiful reflection of 'new fashion']. *XinhuaNet.* http://www.xinhuanet.com/politics/2019-08/20/c_1124898140.htm (accessed 8 October 2019).

14 See the cover page from 18 February 2012 of *Xinmin wanbao* 新民晚报 journal titled: "Yige wei bei lie ru quanshi le ji fenlei shidian fanwei de jumin xiaoqu — lese fenlei lü ruhe da 90% 一个未被列入全市垃圾分类试点范围的居民小区—垃圾分类率如何达90% [How can the garbage classification rate of a residential community not included in the city's garbage classification pilot reach 90%?]", available at http://xmwb.xinmin.cn/resfile/2012-02-18/A01/A01.pdf.

15 Refer to Shanghai Changing District Government (2006) "Shanghai shi wenming xiaoqu chuangjian guanli guiding 上海市文明小区创建管理规定 [Creation of Shanghai Civilised Community Management Regulations]", available at https://www.fengxian.gov.cn/fcz/col3022/20090616/3022-c1965150-f25f-4e14-b360-e5d2afad753a.html (accessed 18 September 2018).

16 Zhou Wang (February 2019) "Why China Is Reactivating Its 'Work-Unit People': The group is viewed by China's leaders as the key to increasing the Party's influence at the grassroots level". *Sixth Tone.* https://www.sixthtone.com/news/1003580/why-china-is-reactivating-its-work-unit-people (accessed 25 October 2019).

17 Original Chinese passage: "*yindao jumin zhudong canyu lese fenlei, zhengfu bumen, shehui duofang liliang dou zai nuli, zhe qizhong, shehui gongyi zuzhi de zuoyong*

buke hushi 引导居民主动参与垃圾分类，政府部门、社会多方力量都在努力，这其中，社会公益组织的作用不可忽视", available at http://www.shio.gov.cn/sh/xwb/n782/n783/u1ai14959_K318.html (accessed 9 May 2018).

18 Refer to Mimi Law (December 2019), "China's Fudan University students in flash mob for freedom of thought". *South China Morning Post*. https://www.scmp.com/news/china/politics/article/3042681/chinas-fudan-university-students-flash-mob-freedom-thought (accessed 12 April 2020).

19 Lily Kuo (July 2019) "'A sort of eco-dictatorship': Shanghai grapples with strict new recycling laws". *The Guardian*. https://www.theguardian.com/world/2019/jul/12/a-sort-of-eco-dictatorship-shanghai-grapples-with-strict-new-recycling-laws (accessed 2 September 2019).

20 The word "participation" is understood as a person, or a group being involved or associated with others in some activity. By citizen participation, we understand the process in which ordinary people take part—on a voluntary or obligatory basis, either alone or as part of a group—and influence a decision involving significant choices that will affect their community.

21 Original passage in Chinese: 在上海做垃圾分类这一块的机构, 主要的经济类源是政府购买服务。所以居民认为ZeroWaste是政府的一部分也是正常的；还有一个原因是ZeroWaste现在开始承接政府项目。所以可以理解.

References

Ackerman, John. 2004. "Co-Governance for Accountability: Beyond 'Exit' and 'Voice'". *World Development* 32 (3): 447–463.

Arantes, Virginie, Can Zou, and Yue Che. 2020. "Coping with Waste: A Government-NGO Collaborative Governance Approach in Shanghai". Journal of Environmental Management 259.

Barker, Derek W. M. 2010. "The Colonization of Civil Society". *Kettering Review* 28 (1): 8–18.

Barker, Derek W. M. 2011. "From Associations to Organizations: Tocqueville, NGOs, and Civic Leadership". In *Alexis de Tocqueville and Democratic Statesmanship*, edited by Brian Danoff and Louie Joseph Hebert, 205–224. Lexington Books.

Bing, Ngeow Chow. 2012. "The Residents' Committee in China's Political System: Democracy, Stability, Mobilization". *Issues & Studies* 48 (2): 72–126.

Blok, Anders. 2014. "Experimenting on Climate Governmentality with Actor-Network Theory". In *Governing the Climate: New Approaches to Rationality, Power and Politics*, edited by J. Stripple and H. Bulkeley, 42–58. Cambridge University Press.

Callon, Michel. 1986a. "Eléments Pour Une Sociologie de La Traduction. La Domestication Des Coquilles Saint-Jacques et Des Marins-Pêcheurs Dans La Baie de Saint-Brieuc". *L'année Sociologique* 36: 169–208.

Callon, Michel. 1986b. "Some Elements of a Sociology of Translation: Domestication of the Scallops and the Fishermen of St Brieuc Bay". *Power, Action and Belief: A New Sociology of Knowledge?*, 196–223.

Dunlop, Alexander. 2019. "Reflections on Authoritarian Populism: Democracy, Technology and Ecological Destruction". *Authoritarianism, Populism and Political Ecology*. https://undisciplinedenvironments.org/2019/02/07/reflections-on-authoritarian-populism-democracy-technology-and-ecological-destruction/.

Ewoh, Andrew I. E., and Melissa Rollins. 2011. "The Role of Environmental NGOs in Chinese Public Policy". *Journal of Global Initiatives: Policy, Pedagogy, Perspective* 6 (1).

Gorman, Patrick. 2016. "Flesh Searches in China". *Asian Survey* 56 (2): 325–347.

Greenhalgh, Susan, and Edwin A. Winckler. 2005. *Governing China's Population: From Leninist to Neoliberal Biopolitics*. Stanford University Press.

Guzman, Gustavo, and Mariana Mayumi P. De Souza. 2018. "Shifting Modes of Governing Municipal Waste—A Sociology of Translation Approach". *Environment and Planning A* 50 (4): 922–938.

Heeks, Richard. 2013. "Development Studies Research and Actor-Network Theory". Actor-Network Theory for Development Working Paper 1. http://hummedia.manchester.ac.uk/institutes/cdi/resources/cdi_ant4d/ANT4DPaper1Heeks.pdf.

Hoffman, Lisa. 2009. "Governmental Rationalities of Environmental City-Building in Contemporary China". In *China's Governmentalities: Governing Change, Changing Government*, edited by Elaine Jeffreys, 107–124. Routledge.

Jeffreys, Elaine. 2006. "Governing Buyers of Sex in the People's Republic of China". *Economy and Society* 35 (4): 571–593.

Kent, Jennifer. 2009. "Individualized Responsibility and Climate Change: 'If Climate Protection Becomes Everyone's Responsibility, Does It End Up Being No-One's?'", *Cosmopolitan Civil Societies Journal* 1 (3).

Latour, Bruno. 1993. *We Have Never Been Modern*. Harvard University Press.

Latour, Bruno. 2004. *Politics of Nature. How to Bring the Sciences into Democracy*. Harvard University Press.

Latour, Bruno. 2005. *Reassembling the Social: An Introduction to Actor-Network-Theory*. Oxford University Press.

Lemke, Thomas. 2001. "'The Birth of Bio-Politics': Michel Foucault's Lecture at the Collège de France on Neo-Liberal Governmentality". *Economy and Society* 30 (2): 190–207.

Lemke, Thomas. 2015. *Foucault, Governmentality, and Critique*. Routledge.

Li, Xiaoliang, Xiaojin Yang, Qi Wei, and Bing Zhang. 2019. "Authoritarian Environmentalism and Environmental Policy Implementation in China". *Resources, Conservation and Recycling* 145: 86–93.

Liu, Chen, and Yijing Jin. 2017. "State of the 3Rs in Asia and the Pacific: The People's Republic of China". United Nations Centre for Regional Development. https://www.uncrd.or.jp/content/documents/5687[Nov%202017]%20China.pdf.

Lourenco, Ramon Fernandes, and Maria Ines Tomael. 2018. "Actor-Network Theory and Cartography of Controversies in Information Science". *Transinformação* 30 (1): 121–140.

Luke, Timothy W. 2016. "Environmental Governmentality". In *The Oxford Handbook of Environmental Political Thought*, 460–474. Oxford University Press.

Pow, Choon-Piew. 2009. *Gated Communities in China: Class, Privilege and the Moral Politics of the Good Life*. Routledge.

Rutherford, Stephanie. 2017. "Environmentality and Green Governmentality". In *International Encyclopedia of Geography: People, the Earth, Environment and Technology: People, the Earth, Environment and Technology*, edited by D. Richardson, N. Castree, M. F. Goodchild, et al., 1–5. Wiley-Blackwell.

Rutland, Ted, and Alex Aylett. 2008. "The Work of Policy: Actor Networks, Governmentality, and Local Action on Climate Change in Portland, Oregon". *Environment and Planning D: Society and Space* 26 (4): 627–646.

Schouten, John W., Diane M. Martin, and Jack S. Tillotson. 2014. "Curbside Cartographies in an Urban Food-Waste Composting Program". In *Waste Management and Sustainable Consumption: Reflections on Consumer Waste*, 102–114. Routledge.

Schulz, Yvan. 2019. "Plastic China: Beyond Waste Imports". In *Dog Days: Made in China Yearbook 2018*. ANU Press.
Shin, Dong Hee. 2016. "Application of Actor-Network Theory to Network Neutrality in Korea: Socio-Ecological Understanding of Network Dynamics". *Telematics and Informatics* 33 (2): 436–451.
Stanforth, Carolyne. 2006. "Using Actor-Network Theory to Analyze e-Government Implementation in Developing Countries". *Information Technologies & International Development* 3 (3): 35.
Swyngedouw, Erik. 2019. *Promises of the Political: Insurgent Cities in a Post-Political Environment*. MIT Press.
Taylan, Ferhat. 2017. "Mesopolitics: Foucault, Environmental Governmentality and the History of the Anthropocene". In *Foucault and the Modern International. The Sciences Po Series in International Relations and Political Economy*, edited by P. Bonditti, D. Bigo, and F. Gros. Palgrave Macmillan.
Thörn, Håkan, and Sebastian Svenberg. 2016. "'We feel the responsibility that you shirk': Movement institutionalization, the Politics of Responsibility and the Case of the Swedish Environmental Movement". *Social Movement Studies* 15 (6): 593–609.
Tsing, Anna Lowenhaupt. 2011. *Friction: An Ethnography of Global Connection*. Princeton University Press.
Wan, Xiaoyuan. 2015. "Governmentalities in Everyday Practices: The Dynamic of Urban Neighbourhood Governance in China". *Urban Studies*. 53 (11): 2330–2346.
Wang, Alex L. 2017. "Explaining Environmental Information Disclosure in China". *Ecology LQ* 44: 865.
Wang, Jinjun, and Qun Wang. 2018. "Social Autonomy and Political Integration: Two Policy Approaches to the Government-Nonprofit Relationship since the 18th National Congress of the Communist Party of China". In *Nonprofit Policy Forum*. Vol. 9. De Gruyter.
Wang, Meiqin. 2019. "Waste, Pollution, and Environmental Activism: Wang Jiuliang and the Power of Documenting". In *Socially Engaged Art in Contemporary China*, 53–81. Routledge
Woldring, Henk E. S. 1998. "State and Civil Society in the Political Philosophy of Alexis de Tocqueville". *Voluntas* 9 (4): 363–373.
Wong, Linda, and Bernard Poon. 2005. "From Serving Neighbors to Recontrolling Urban Society: The Transformation of China's Community Policy". *China Information* 19 (3): 413–442.
Yonay, Yuval P. 1994. "When Black Boxes Clash: Competing Ideas of What Science Is in Economics, 1924–39". *Social Studies of Science* 24 (1): 39–80.
Zhang, Huxiang, Kaiming Wu, and Hui Song. 2019. "Qingnian shehui zuzhi canyu jiceng shehui zhili shijian lujing yanjiu—yi ai fen huanbao tuijin shequ lese fenlei de shijian wei li 青年社会组织参与基层社会治理实践路径研究——以爱芬环保推进社区垃圾分类的实践为例 [Research on the Path of Youth Social Organizations Participating in Grassroots Social Governance——Based on the Practice of Aifen Environmental Protection NGO]". *Qingnian xuebao* 青年学报 [Youth Research].
Zhang, Xuefan, Jing Wang, and Li Xu. 2019. "Between Autonomy and Supervision: The Interpretation of Community Supervisory Committee Reform in Hangzhou, China". *Cities* 88: 91–99.
Zhao, Litao. 2019. "The Party in Grassroots Governance". In *The Chinese Communist Party in Action: Consolidating Party Rule*, edited by Yongnian Zheng and Lance L. P. Gore, 288. Routledge.

Zhou, Ming-Hui, Shui-Long Shen, Ye-Shuang Xu, and An-Nan Zhou. 2019. "New Policy and Implementation of Municipal Solid Waste Classification in Shanghai, China". *International Journal of Environmental Research and Public Health* 16 (17): 3099.

Zimmer, A. 2010. "Urban Political Ecology: Theoretical Concepts, Challenges, and Suggested Future Directions". *Erdkunde* 64 (4): 343–354.

5 Embracing the market

Introduction

In 2011, Song (not his real name), formerly an English teacher and real-estate consultant in Shanghai, returned to his hometown, Hunan, to become a farmer. His story embodies the recent wave of "new farmers" (in Chinese, *xin nongmin* 新农民, *xin nong ren* 新农人, *xin nongfu* 新农夫) leaving their "comfortable" jobs in the city to devote themselves to sustainable farming (Scott et al. 2018). Song has not had a smooth experience, however. Aiming to strengthen rural communities and promote sustainable agriculture in his village, in 2012, Song created Farming, a social non-profit organization (*shehui tuanti* 社会团体). Yet after facing several hurdles, Song later decided to adopt a for-profit model, a social enterprise. Since 2016, following the restrictive measures for the establishment of NGOs, many Chinese and foreign SGOs have embraced this strategy.

Moving beyond a state-centric approach, in this chapter, I will focus on the rise of social entrepreneurs in Shanghai. Following on from Chapter 4—where a core focus was to decipher SGOs' roles as allies of the state—here I will explore the motivations behind the growing marketization of SGOs. Is this trend a "tactic" (de Certeau 1984)[1] used to circumvent the coercive pressures and limitations described in the previous chapters? Indeed, by adopting market strategies and developing more independently from the state, SGOs should, logically, be able to strengthen their agency. Yet, as I will show through the case of Farming (and other organizations observed in Shanghai), their motivations and outcomes are also being challenged by new hurdles.

Shanghai represents a unique case study to examine the development of civil society in China's new cosmopolitan areas. First occupied by the British in the 18th century (following the First Sino-Japanese War of 1894–1895), then quickly followed by the French and the Americans, the city has grown, from the beginning, as a fractured settlement. In essence, Shanghai was two cities, one a Chinese city under the authority of the Chinese government, the other an international settlement consisting of concessions controlled by foreign powers. Such a strong international presence in the city made Shanghai

DOI: 10.4324/9781003231325-5

the world's third most important banking capital by the end of the 19th century (after New York and London), and, as such, a symbol of China's exponential growth and international openness. Deng Xiaoping (China's paramount leader after the death of Mao Zedong in 1976) himself christened the city the "Head of the Dragon", reinforcing the city's role as China's economic hub. Given Shanghai's exposure to the outside world it is not surprising that the city became a testing ground and a showcase for the country's ongoing efforts to reduce air pollution and control urbanization (I will further explore this point in the next chapter). Yet there is probably also no city that better epitomizes the trend of social policy decentralization and how this has been parallel to economic marketization.

In this chapter, I explore the growing presence of social marketization in Shanghai through the lens of social enterprises. These have been developing considerably around the globe and embody one of the many categories of SGOs operating in China (refer to Chapter 2). Yet they represent a considerable new trend as 50 percent of them were founded between 2013 and 2015.[2] But does this increase represent a logical response to the CCP's efforts to tighten control over civil society? Or should we develop a more nuanced understanding of the organizations' development, roles, and purposes?

By focusing on the emergence of social enterprises, I aim to question whether their growing presence is a strategic development model, for both foreign and Chinese SGOs, to expand their social influence and escape from the hands of the state; or if it is rather part of a global trend. The chapter will mostly revolve around the following question: Why do social enterprises represent such an appealing option, in comparison with traditional NGOs, to respond to emerging socio-environmental issues in Shanghai? I will focus in particular on the economic, social, and political context of Shanghai and the specific opportunities and constraints that it creates. As explored before, in Shanghai, SGOs develop in close alignment with the municipal government apparatus. But does the important role of the market economy in Shanghai open extra spaces for SGOs to develop their agency?

To respond to these questions, this chapter is divided into three parts. The first begins with an overview of the literature on social entrepreneurship. I show how, in an authoritarian environment like China, SGOs are simultaneously shaped by the respective influences of state institutions, market actors, and citizens. In the second part, I then draw on Actor-Network Theory (ANT) to present an original case study: the "story" of Farming. Building on the previous chapter, I assess Farming's network expansion, specifically in trans-local cooperation. I analyze the networks and alliances used by Farming to upscale its goals and why these led the organization to marketize. Lastly, to further reflect on the phenomenon of marketization, I continue exploring the concept of *governmentality*. Adopting a Foucauldian perspective, I show how embracing market strategies leads SGOs towards depoliticized forms of conduct.

Social enterprises

Various definitions for social enterprises have been put forward in the literature (Tauber 2019). Emerging from hundreds of years of development of charitable organizations in Western countries, it is a renewed concept that uses business as an instrument for social development (Dart 2004). Coined as the future of social welfare, the number of social enterprises has been steadily increasing since the Asian financial crisis in the late 1990s and the global financial crisis of 2008 (Kim and Defourny 2011; Poon 2011; FYSE 2012; M. Zhao 2012; Wang et al. 2015; Oberoi et al. 2019).

Although social enterprises have been thriving since the 1990s, a single, clear-cut definition is lacking in the literature. Closely related to the concept of "social economy", the description remains "fuzzy" (Huybrechts et al. 2016). Because an organization's legal form depends on the country where it is based, the term "social enterprise" is used in different ways in many parts of the world. The EMES (*l'Emergence de l'Entreprise Sociale en Europe*) has been attempting to create an "ideal type" since 1996.[3] As a response, the European Commission published a description including three criteria: social objectives, limited profit distribution, and participatory governance. In China, several experts and practitioners are trying to create a definition which could better reflect and adapt the concept to the Chinese context.

But given the lack of a universally accepted definition and their close relation to charitable organizations, the terms social enterprise and NGO are often misused. The key difference between an NGO and a social enterprise is the revenue model. Traditional organizations rely primarily on charitable contributions, public funding, and foundation grants to support their actions. Social enterprises develop business models that produce a profit which is later reinvested to create social value. Other qualities such as innovation, autonomy in leadership, and decision-making distinguish social enterprises from other businesses and organizations (Tauber 2019). In the literature, scholars refer to the rising numbers of social enterprises as a strategy for social actors to respond to funding and volunteering difficulties (Pallotta 2010). Still, the list of features mentioned here is not exhaustive and social enterprises can take many forms: some are created, operated and/or owned by non-profit, charitable organizations to generate income and/or to further their social mission.

Social enterprises in China

In China, practitioners introduced the concept of social enterprises at the beginning of the 2000s through several papers and conferences (Zhao 2014).[4] This gained particular attention in 2006, following the translation of two internationally bestselling books: *How to Change the World* by David Bornstein, and *The Rise of the Social Entrepreneur* by Charles Leadbeater; and then, subsequently, in 2008, as companies helped relief efforts for the Sichuan earthquake (Hsu 2017; Zhang and Shi 2017). Since 2009, the British Council has influenced the sector significantly by offering specific skills training, mentoring, and access to UK expertise. By 2017, for example, a training program called "Skills for Social Entrepreneurs" had trained over 3,200 social entrepreneurs and pledged nearly 37 million RMB to 117 social enterprises.[5]

But, unlike in the UK, in China there is no specific legal form for social enterprises and as a result, organizations operate under diverse operational models. According to the Foundation for Youth Social Entrepreneurship (FYSE),[6] organizations take on two registered forms in China: NGOs, or business entities. Of the 102 social enterprises interviewed by Seforïs (Huysentruyt 2016), some social enterprises were registered as a business (*gongsi* 公司), while most were registered as limited companies. Only one-third of the organizations were registered as civil non-enterprise units (*minban fei qiye danwei* 民办非企业单位), the Chinese legal form for NGOs (Z. Zhao 2014, 13). As for the organizations in our sample (refer to Appendix C), these mostly had a traditional grassroots charity background. Otherwise, many were start-ups, either choosing the social enterprise path for convenience (e.g., registration hurdles), or ideology. The following proverb (which I heard on various occasions during my fieldwork) embodies their main "ideological" stance: *Give a man a fish, and you feed him for a day. Teach a man to fish, and you feed him for a lifetime.* Organizations used this to position themselves in contrast to traditional philanthropic methods, perceived as a somewhat old-fashioned, "candy giving" approach.

Despite the considerable growth of social enterprises, the Chinese government has not yet established a clear legal position or relevant legislation. The social welfare enterprise model (*shehui fuli qiye* 社会福利企业), in place since 1949, represents the approach with the closest resemblance to the social enterprises found in Europe (Zhou et al. 2013).[7] They provide employment opportunities to disadvantaged groups, notably those with physical handicaps. In Zhao Meng's words, this type of organization is "a for-profit enterprise that receives social welfare status from the government and hence enjoys tax benefits, government loans, and other kinds of support" (2012). Among the diverse forms of SGOs currently operating in China, social welfare enterprises represent one sub-category. They are business enterprises that are registered under the Ministry of Industry and Commerce but that need to maintain an administrative relationship with their supervisory government agency. For over 65 years, they acted as key players in the development of jobs for people with disabilities. More recently, the government forced them to operate under the double constraints of regulations *vis-à-vis* disabled people and welfare enterprises, meaning that their number has dropped since 2007. Their operational model contrasts with the organizations under study because they rely on government support, and therefore don't enjoy the same level of autonomy.

What is a Chinese social enterprise?

The social enterprises in our sample (refer to Appendix C) contrast with social welfare enterprises because they do not rely on the government for support. Despite a lack of legal recognition, several practitioners and scholars have tried to develop a definition that is specific to the Chinese context. According to the presentations I attended in Shanghai between 2016 and 2018, China's social enterprises are defined by three elements: (1) achieving social good as the primary goal; (2) using business operation methods; and (3) following a profit-sharing model. The Social Enterprises in Greater China Report, for instance, stresses that social enterprises should "solve social problems as the primary goal of their organization, rather than achieve a profit (*yi jiejue shehui wenti wei jigou de shouyao mubiao, er fei yi yingli wei shouyao mubiao mùbiāo* 以解决社会问题为机构的首要目标 而非以营利为首要目标)" (Zhang and Shi 2017, 9).

The Non-Profit Incubator (NPI), highlighted in Chapter 3, has been particularly active in advancing the concept. NPI is a civil non-enterprise unit—referring to public institutions, social groups, and other social organizations engaged in non-profit activities—but nurtures both NGOs and social enterprises. According to Ding Li, its vice-president, social enterprises are self-sustained organizations that create a social impact and need to have independent regulations and a board of directors.[8] Several requirements need to be respected. In contrast to a traditional NGO which has zero financial return, or a commercial business that has (in general) a financial return of 25 percent or more, the financial return of a social enterprise should account for 0–10 percent of the organization's total profit.

Even though there is no unified voice on the topic, all commentators agreed that social enterprises should not follow profit-maximization as their principal goal. For Jiang Jiawei 张嘉伟 (Interview 18 June 2018), General Secretary of the Social Enterprise Research Centre (*shehui qiye yanjiu zhongxin* 社会企业研究中心), social enterprises distinguish themselves from NGOs because of their flexibility (status, resources, and activities), and profitability. The features highlighted by practitioners in the field are also discussed in the Chinese literature. Wu (2018) states that social enterprises are based on "the innovative use of market tools to solve social problems (*pingjie zai chuangxin xing yunyong shichang shouduan jiejue shehui wenti* 凭借在创新型运用市场手段解决社会问题)" and represent "a mixture of commercial enterprises and non-profit organizations (*shangye qiye he fei yingli zuzhi de hunhe ti* 商业企业和非营利组织的混合体)". Yet blurring the lines between business and philanthropy creates ambiguities. As the Communication Manager of NPI stressed (Interview October 2016):

> Well, some of them are registered as business and others as NGOs. We have two different systems: industry and commercial administration and civil affairs administration. There are two separate sectors. Sometimes

> they just register on this one or that one. So, there is no legal definition. They just define themselves: OK, I think I am a social enterprise.

In the same light, a professor at Tsinghua University (Interview December 2016) argued that it took years of interactions and joint learning processes between the central/local government and NGOs to achieve a legal system capable of regulating the sector. The same is likely to happen to social enterprises, he said. According to this professor, the lack of official legal recognition leads organizations to become "social enterprises" because it enables them to stay in a grey area like those non-profit SGOs that, for nearly 30 years, either registered as for-profits or didn't register at all. Meanwhile, this increase has also been due to other institutional, political, and economic factors.

An emerging trend in Shanghai

There are as many reasons for organizations to adopt a social enterprise model as there are different types of organization currently active in the Metropole. With a long history of foreign settlements and cosmopolitan culture, Shanghai has developed certain conditions which are favorable to their growth. A first point to take into consideration is the fact that, contrary to Beijing, Shanghai receives far less political and cultural influence from the central authorities. The city is home to an important number of major Social Impact Investment Institutions: Transi.st, Narada Foundation, More Love Foundation, Yu Venture Philanthropy, Dianzan, and many more. As a global center for finance, innovation, transportation, and commerce, Shanghai has also produced a large pool of potential social entrepreneurs, support institutions (e.g., Social Enterprise Research Centre), and forward-thinking governments. The city's reputation as a leading global innovation hub also attracts a steady flow of foreigners,[9] and Shanghai is China's most popular city for expatriates and international businesses. For the fourth time in six years, it topped the list of the ten most attractive Chinese cities for expatriates.[10]

Moreover, Shanghai also profits from a high level of foreign direct investment. Even though there was a sharp decline in investments following Trump's anti-globalization politics and Xi's focus on national markets, Tesla's new Shanghai Gigafactory represents "the most eye-catching success" of Xi's new trade and investment liberalization approach.[11] Tesla enjoyed considerable benefits in building the first wholly foreign-owned car plant in China. The city is also investing nearly 100 billion RMB over five years into a fund to support the development of start-ups, business infrastructure, and even to attract high-skilled workers to the area (while, at the same time, it seeks to limit access to "undesirable" foreign workers, as I will explain in Chapter 6).[12] Other measures taken by the city include trade and investment liberalization, market facilitation policies, intellectual property rights, the banning of forced technology transfers, and equal treatment for foreign firms in government procurement. These measures are being developed in response to the Foreign Investment Law, which took effect on 1 January 2020.[13] Shanghai has been implementing its own set of measures to improve its business environment.[14] It is not rare to see policies first being piloted in the city's free trade zone and later expanded to the municipality level. All these factors create a favorable environment for western companies to develop in the mega-city.

Meanwhile, registering as a new business entity in China got easier. Free trade zones, for instance, offer numerous advantages—the possibility of using a virtual office as an official address or advantages for tax return—to foreign investors. International companies that are interested in investing put strategic procedures in place to register as Wholly Foreign-Owned Enterprises, also called WFOEs. A WFOE is a limited liability company that allows entrepreneurs to engage in profit-making activities and hire both local and foreign employees. Several of the social entrepreneurs in our sample were taking advantage of these new "conveniences" to enter the market. As the founder of Turtle, a SGO focused on environmental advocacy, argued (Interview April 2016):

> [W]e are going to have this meeting tomorrow to set up an organization in China. But again, not as a non-profit, because it is almost impossible, but as a WFOE. Basically, if you want to set up a company in China, you have two options: one is that you can do it as a Chinese company, so you need to have a Chinese person to register, and your capital and economy must be really small; or you register as a Wholly Foreign-Owned Enterprise (WFOE): so, as a foreigner, I can set up a company. But, of course, there are different implications. So, you know, now that we have managed to have some money from Turtle from these last two years, I can use it... before, setting up a WFOE was very expensive, time-consuming, and so on. But now, specifically, as an entrepreneur and environmental-focused worker, the government has made it much cheaper. I also met a really good lawyer two weeks ago who gave me a very good guideline on how I can do it at less cost. So, now I closed the Hong Kong company and I'm setting up Turtle as a company in China. In the future, as a social enterprise, because it is kind of impossible to do "business" in China without having a local setup.

For Turtle and other organizations (e.g., Pig, Wolf and Panda; see Appendix C), establishing a WFOE legal entity became a strategic way to survive. Although they are non-profit in nature, they preyed on the "elastic" feature of the social enterprise concept to bypass the Overseas NGO Management Law. As one social entrepreneur pointed out (Interview April 2017), they only need to change their official name from NGO to social enterprise should any problem emerge. One might therefore imagine that SGOs are embracing a social enterprise to respond to these hurdles, however that would be a far too simplistic answer.

Based on the current tax regulations, it can be more attractive for foreign businesses to adopt a for-profit status compared to register a permanent and legal office as an NGO. As Hogan Lovells (2016) explains, while a WFOE that is not making profit pays no enterprise income tax, a representative office that is not making profit but is incurring expenses will still have to pay enterprise income taxes.[15] Moreover, a for-profit status can have certain beneficial tax implications for SGOs:

> With the mandatory registration requirement, it will be possible for the Chinese authorities to track every movement and activity of Foreign NGOs in China, which can lead to significant tax exposure for Foreign NGOs in China if their activities are not planned properly. Also, compared to other types of legal entities in China, ROs (representative offices) are subject to a significantly higher tax burden, so Foreign NGOs will want to be fully aware of these tax exposures before undertaking any activities or establishing ROs in China.
>
> (Hogan Lovells 2016, 6)

Some SGOs adopt a hybrid model by registering as two entities: a company and an NGO[16], as, for example, in the case of NPI. As before mentioned, NPI works closely with the government but there are state restrictions on which organizations can be included in the official incubator program. According to NPI's Marketing and Communication Manager (Interview October 2016), this stimulated NPI to start the "social enterprise accelerating program". By establishing a for-profit organization (social enterprise), they create revenue which enables them to become more independent from government resources and, therefore, incubate organizations without receiving the government's approval. Yet they also continue to work closely with governmental agencies. Thus, this change of strategy cannot be interpreted as a direct response to governmental pressures. It could also be seen, according to both the Marketing and Communication Manager and the Program Manager of NPI (Interview October 2016; Interview December 2016), as a way to implement new strategies while reducing the organization's dependence on state support. Interviewees depicted the government as conservative and not open to innovative and risky strategies, and they argued that this was an important factor in their deciding to develop social marketization strategies.

Finally, a social enterprise model is also seen as a good approach to help SGOs move beyond the unstable and unskilled workforce as typically found in NGOs. Unlike grants, government contracts forbid the use of funding for salaries and/or restrict administrative costs to 10 percent of the total value of the contract (Interview 6 May 2016). By becoming financially independent as a social enterprise, organizations can increase the salaries they offer and attract new, highly skilled talent. This is extremely important, and many also see social enterprises as a means of recalibrating gender inequalities and increasing the attractiveness of the social sector. By becoming more financially independent, organizations are able to invest new resources and increase workers' salaries. As Li and Merkel (2017) explain:

> Low salaries and high living costs present another challenge, making it difficult if not impossible for men to consider working in the social sector if they have to support a family. Even though the majority of Chinese women living in urban areas have a job, men are still expected to be the breadwinners. Based on recent statistics released by the Catalyst, women often receive only 70 percent of the salary men receive for doing the same job. All this contributes to the fact that 70 percent of the employees in the social sector are young women, whose employment does not have a high status in society […] NGOs and early-stage social entrepreneurs must often deal with annual turnover rates as high as 40 percent, which means they must constantly engage in costly recruiting efforts, manage an unstable workforce, and struggle to retain their culture.

So, adopting a social enterprise can point to organizations having very different political and/or economic motivations. Although the various pressures from the top should not be minimized, it is important to highlight other factors that drive SGOs to choose this path of development. Nonetheless, adopting this blurry and unrecognized social enterprise model is not a magic wand for all, as we will see in the following section.

An open door, but not for everyone

For those working on sensitive issues such as human rights, religion, or LGBTIQ topics, it is much harder to adopt a social enterprise model. I previously stressed that most social enterprises come from the non-profit sector (FYSE 2012, 25). But not all SGOs can shift towards a viable for-profit model. Because it works on gender issues, the organization "Panda", for instance, overcame the fact that it could not register as an NGO by establishing itself as a social enterprise. To survive, the organization mainly focuses on developing its activities on social media and sells merchandise (bags, pins, etc.) to make a profit. Most of the SGOs analyzed here were selling services to the private sector while four of them were trying to find solutions to scale their business model. Generally, their ability to start a business depended on the founder's capital or personal resources (*guanxi* 关系), and on sponsorship, crowdfunding actions (e.g., Shanghai Soup, see Box 5.1), or donations.

Box 5.1. Micro funding with Shanghai Soup.

Shanghai Soup is a project based on the successful Detroit Soup project. The soup served at each event (held at various locations in central Shanghai) is made from food provided by local suppliers and food from restaurants that would have been otherwise wasted through overstocking or cosmetic damage. Participants pay 50 RMB for soup and vote on which project they think would benefit the community the most. The winner goes home with the money raised at the door and uses it to carry out their project. This initiative was recognized by social entrepreneurs in the field as helping Shanghai residents and social entrepreneurs to connect, and share ideas and community resources.

In the following section, I further explore how, under these conditions, marketization is becoming increasingly intertwined within SGOs' development mechanisms. In line with Carrillo and Duckett (2011), I stress that SGOs often organize national state programs in line with their specific economic and social development. To prove how various hurdles and specificities of Shanghai lead SGOs to increasingly engage with the market, I develop an actor-network analysis of Farming's urban–rural bridging program. The analysis has two goals: (1) to identify how embracing market strategies enables SGOs to forge networks and alliances to "upscale" and "out-scale" their struggles; and (2) to show why this approach reinforces rather than weakens the "green" consensus thesis advanced in the book.

Scale here is used to refer to the social relations that connect local human and non-human actants to broader political, economic, cultural, and ecological scales. These scales are "neither natural nor fixed but are produced through frictions between social practice, environmental processes and structural forces" (Boelens et al. 2016). Concepts such as "sustainability", "environmental protection", and "ecology" give form to an array of interconnected issues where it becomes possible to frame a discourse and/or engage across one or multiple scales (Swyngedouw et al. 2014).

Farming: growing organic connections

Although the promotion of ecological development led to a greater emphasis on food policy by China's leaders of Party and State, decades of frequent food scandals have damaged public confidence in the country's food safety regulatory system. This situation led to the emergence of a new class of Chinese farmers, such as Mr. Song, who choose to dedicate themselves to fair and sustainable agriculture. After returning to Hunan, Song quickly realized the necessity of bringing the complexities of the rural–urban divide into the city sphere. Since food is becoming a hot political issue, it was the perfect moment to catch the interest of new actors and, thus, upscale the struggles and interests of farmers to get the attention of city-dwellers.

The inability of the party-state to respond to consumer anxieties coupled with the expansion of China's highly educated middle class are the main drivers behind the emergence of small-scale ecological farmers all around the country (Ding et al. 2018). In her study on changing consumer culture,

Leggett (2017) stresses that this situation creates new opportunities for civil society organizations to fill state voids. As she argues, farmers have access to a multitude of tools (e.g., social media) which they used to connect, establish, and maintain a relation of trust with consumers, even if they are established in remote areas. Scott et al. (2018, 302) rationalize the emergence of such groups as a site for "nascent activists deploying grassroots community organizing strategies".

Yet do these initiatives create spaces for civil society emancipation? As I explore below, Song's organization Farming went through several incarnations. Starting as a social organization—one of the three legal recognized forms for non-governmental organizations in the Mainland—later shifting to a for-profit model, it continuously adapted to respond to different limitations and upscale its goals. The analysis below is based on one semi-structured interview and several informal conversations with Song. I also actively participated in several of their events and observed their social media content by following their public messages and interactions with clients and supporters in private WeChat groups.

Bridging the urban–rural divide

The goal of the organization Farming is simple: linking urban families with rural farmers, while providing traceable and healthy food to city-dwellers. As Song explained to me, city families sponsor a rural family and in return they receive a weekly delivery of food products, free of harmful chemicals and genetic modifications, as he proudly explained. Through this project, Song aimed to develop a small-scale, healthy, green, and non-profit agriculture—by getting rid of the "middle men" (as Song referred to supermarkets, retailers, and distribution chains) between farmers and consumers. As he said, this method would help to strengthen the precarious situation of farmers in the countryside. Unlike other trendy agricultural movements in the city at the time, Farming didn't aim to bring farmers closer to city-dwellers or to "green" the city, as for Song, sustainable agriculture is incompatible with urbanization. Food needs to grow far away from pollution, in places where "there are no factories, mines or any highways at all, so the environment is free from any industrial pollution or car emission" (Interview November 2016).

The work of Farming goes against the growing tendency of urbanizing agriculture stressed in Ding et al.'s (2018, 81) analysis, and what they term as the new urban agricultural geography of Shanghai, meaning, "a new spatial location for small, mainly family, farms in the city and its suburbs, allied with the emergence of new farmers with motivations associated with ecological farming and the development of 'activist' networks of customers". Song's main objective is to confront the income inequality between China's rural and urban areas by shining a spotlight on the political, economic, socio-cultural, environmental, and psychological problems that have surged in recent years.

Taking advantage of his English skills, Song engaged in different strategies by reaching out to national and international organizations such as the Slow Food Movement or Wayne Weiseman's world-famous Permaculture Project.[17] It was by building up networks and alliances with a diversity of actors that Song was able to expand Farming's network. Yet the journey wasn't easy, and Song had to face numerous hurdles that would lead the organization to marketize throughout the years.

It could be tempting to analyze the marketization of SGOs as proof of emancipation. Yet here I rather want to argue that by embracing market strategies, organizations risk falling into ambivalent and apolitical narratives. I use ANT—or Actor-Network Theory—to show how the ideologies and practices of Farming slowly became dissociated from the first political stances regarding the dangerous level of poverty and inequality in rural areas. By sharing images of his farmers' entrepreneurial skills with the wider public, Farming obliterated the perceptions of inequality in favor of portraying a model of "sustainable urbanism" and "rural vitalization" which reinforces the goals of the party-state. To strengthen this argument, I will further highlight why and how the movement expanded from 2011 to 2018, and then continue by analyzing Farming's key moves and players.

Problematization

Phase I: establishing common interests

The *problematization* phase started in 2008 as soon as Song returned to his hometown and established his farm, *Song's Organic Farming* (fictitious name). Upon his arrival, he quickly experienced the effects of the rural–urban gap. As he explained:

> Since 2008 … 80 percent population of my village just moved to the cities for a better life. Only 100 out of 900 people have remained. Now, most people staying in the village are over 70 years or just below 10 years old.[18]

For Song, the depressing environment of rural villages is even worse when we consider the left-behind children who suffer the most from the situation.[19] In Chapter 4, we saw that *problematization* refers to the way in which an actor represents a specific problem (either a group or an individual). It is the first phase of the *translation* process that helps to describe the set of alliances or associations created between actors and the consequent establishment of dependency among them (Latour 1996).

Song was aware about farmers' precarious conditions as he had experienced them himself after returning to his hometown. Dismayed by the situation of farmers, Song created Farming. Having long experienced city life, he knew that city-dwellers were eager to get access to healthy food. But he was fast confronted not only with a lack of an organizational structure but also a lack of knowledge of ecologically sustainable food practices among his fellow farmers. Another difficulty was the farmers' complete disconnection from the market that created considerable challenges when it later came to launching an autonomous supply chain. As Song explained to me, an individual farmer cannot harvest enough products and sell them directly to the market. Besides, farmers' households are separated from one another, making it hard to follow specific regulations and, thus, establish strict quality control mechanisms.

Aware that a lack of an organizational structure was hindering their development capacities, Song started by getting a People's Republic China Remarks Organization Code Certificate (*zhonghua renmin gongheguo zuzhi jigou daima zheng* 中华人民共和国组织机构代码证) in 2012. An organization code is indispensable for carrying out tasks such as opening a bank account or getting an export license. Although having the code helped deal with quality control, it didn't have an effect on another fundamental struggle: competition. As commercial organizations mainly focus on profit, they mainly compete against the farmers to set low prices. In such conditions, it became extremely difficult for Farming to survive.

Three years after returning to Hunan, in 2015, Song created a cooperative. Under a cooperative umbrella, farmers could better coordinate themselves. This improved the communication among farmers and streamlined spaces to discuss norms, customer relations, or logistics. According to Song, having a cooperative solved the difficulties of Hunan farmers and opened new opportunities to *enroll* them in his project: create ecologically sustainable farming practices, with no use of chemicals or any other harmful substances and establish a relation of interdependence with city-dwellers. Yet, as Song stressed, persuading his fellow farmers to give up pesticides was not a simple task. Song needed to convince farmers it would enable them to make higher revenues.

To this end, it became indispensable for Song to *enroll* supporters in the city who would agree to pay more for their healthy products. Far away from Shanghai, it was crucial to develop a tool to connect farmers with city-dwellers. The solution was to create an application, not only to process orders but also to maintain a relationship of trust, which is how Song came to develop Farming into an urban–rural bridge program. Farming rapidly became the *obligatory passage point* (see Chapter 4, p. 85–86). Figure 5.1 succinctly describes Farming's program and also identifies its *secondary obligatory passage points*, as well as the actors linked to them.

Phase II: extending scale

Learning from its first attempt to do business as a small private enterprise, in 2015, Farming officially registered as a social organization (*shehui tuanti* 社会团体). By acquiring this status, Farming officially became a Chinese rural cooperative. In practice, in China farmers' cooperatives are an essential way

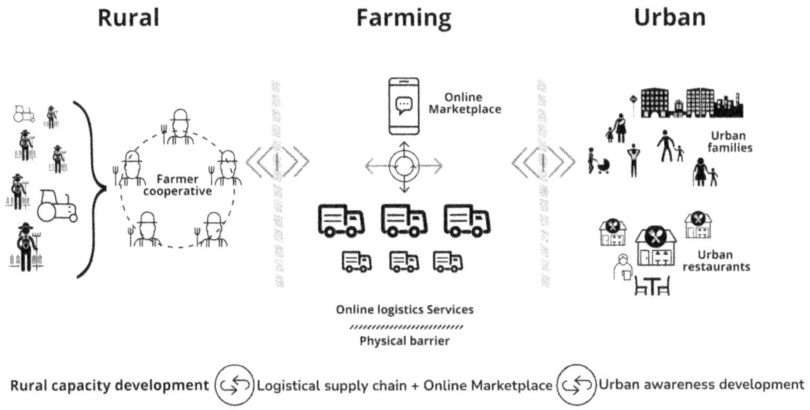

Figure 5.1. Farming program framework. Source: Adapted from one of the co-founder's presentations attended by the author in Shanghai.

to organize scattered farmers, improve agricultural production, and facilitate their access to the market (Huang 2014). Being a cooperative also eases economic cooperation and market integration (Song et al. 2014). Yet, as Yu Keping (2011, 87) stresses: "institutional deficiencies invariably result in leaving many civil organizations at a loss as to what to do". As Yu explains, farmers' cooperatives have suffered from institutional difficulties because they do not enjoy a clear legal position. They therefore have to deal with many legal, institutional and taxation constraints. For example, a cooperative is subject to tax, while sales made by an individual farmer are not. Moreover, banks do not recognize the cooperative as legal persons and will not grant loans. As Yu (*ibid.*) points out, "This is without a doubt not beneficial for expanding the scale of rural professional cooperative organizations".

So, even though Song created a cooperative which allowed him to disseminate the information and knowledge about Farming, many difficulties remained. A cooperative structure was essential to accumulate experience and maximize its members' influence but, again, this was not a good strategy to face competitive markets, and this led Song to recognize that registering as a social organization was not efficient. As he explained in his blog (on Weibo), consumers were not yet willing to pay more for sustainable food, and thus it was challenging to survive in a competitive market and rely on scarce and unreliable public funds. It was also expensive to get an official certificate endorsing the fact that your food is organic. Farmers could not deal with all these constraints, and these challenges were also not solved when Farming adopted the non-profit label.

Phase III: aiming for a self-sufficient economy

Farmers all around China are deploying efforts to establish novel forms of collective action in the hopes of achieving a better life in an environment that is increasingly competitive (Song et al. 2014). The cooperative model, as explained above, is essential to organize local farmers because it helps them to combine their efforts and work together on a common, easy to manage piece of land. A cooperative status ensures that the farmers own the land and a share of any extra profits. Yet, by the end of 2016, Farming changed its strategy and became a social enterprise. For Song, spending time asking for resources is not a good solution if you want to increase your livelihood as a farmer, explaining that "An 'NGO' model is not appropriate". The goal of the organization is to meet farmers' economic needs by providing a service that benefits consumers. Its primary activity is to make money to empower rural farmers, regardless of whether farmers did it sustainably or not.

The paragraph above outlines Song's decision to switch the focus of Farming from organizing farmers under a cooperative umbrella to functioning as a farmer-city-consumer bridge platform. Since the relaunch of Farming in late 2016, farmers continue to be organized as a farmers' cooperative (social organization) while they sell their products via Farming, which is, theoretically, a social enterprise. It was at this specific point in Farming's history that I met Song. He was certain that a social enterprise model would solve his previous problems. Even though Farming was not officially registered as a social enterprise, he kept presenting the organization in that way, and slowly, Farming developed into the practice of community-supported agriculture as a social enterprise model. Later, in 2018, Farming was registered a social enterprise in Hong Kong—where Farming is now legally recognized and supported—in cooperation with foreign colleagues and under a different name. On their website, the Hong Kong organization is presented as operating under its Shanghai-based sister organization, Farming.

As highlighted by one of the co-founders of Farming's Hong Kong social enterprise (Informal interview August 2018), by registering in Hong Kong, the organization gains access to foreign capital. This move was not specific to this organization, as similar strategies were used by other SGOs observed in the field, all with the aim of countering the Charity law. As stated before, the law restricts non-profit organizations from obtaining overseas funding. By registering in Hong Kong, Farming therefore gained greater access to foreign funds.

Interessement/enrollment

As explained before, Song captured the interest of other farmers by providing them with an opportunity to access new market channels. Besides, by creating a cooperative, Song shaped a framework within which he could share his ideas of sustainability and reinforce his relations with local farmers. Because of his expertise, he is recognized as the farmers' *spokesman*, rendering himself indispensable in the network (Latour 2004).

Meanwhile, Farming expanded its activity beyond the farmers' environment to different scales and places by maintaining extra-rural and global relations. These scaling strategies empowered Song to reach urban consumers without needing large vertically- or horizontally-integrated corporate structures. He engaged with a multiplicity of actors, from city consumers, restaurants to environmental activists by taking part in Shanghai's environmental activities.

During events run by these activists, the bucolic view of peasant life was presented to emphasize a traditional way of the organization Farming. Images of the beautiful countryside were shown to Shanghai's rising middle-class who were craving food security and a reconnection with nature. WeChat in particular was an important tool and *mediator* in this strategy (see Chapter 4, p. 87), and technology made it possible to scale this project. WeChat and Farming's Online Marketplace allowed consumers to buy and pay for food online, and maintain a close dialogue with farmers in the countryside. Because Hunan is far from Shanghai, WeChat is indispensable to the stability of the network, used to ask questions, and share videos and pictures of the products. During one interview, Song showed me a photo of a river that was posted in one of those groups. Below the photo, a farmer explained in a comment that the fish he was selling came from that same river. WeChat is used to reassure consumers and respond to the anxiety of Chinese consumers after years of food scandals.

Broadly speaking, Song's motivation for changing the food system is based on ideas of natural, traditional, and traceable healthy food shaped by a network of alliances between farmers and city consumers. Farming creates an ecologically sustainable community by sharing knowledge with farmers. Farmers' associations or cooperatives activate the farmers' resources (land, labor, and housing) while Farming helps them build up a group identity through the Farming platform. This identity is appealing to many of the urban citizens, and allays many of their concerns.

Yet, Song also reinforces his organization's network by involving international actors such as the Slow Food movement.[20] I witnessed firsthand how he grabs the attention of clients when I took part in a dinner co-organized by Farming, Slow Food, and a high-end vegan restaurant in Shanghai. They invited a number of Chinese and foreign guests to taste different recipes cooked with Farming products. Song put a lot of effort to attract foreign clients, as reaching an international public was a major strategy for him (Informal interview during the dinner). According to Song, an international image is very attractive to Chinese consumers. All of Farming's online posts and marketplace were bilingual in English and Mandarin, with repeated use of images of foreigners visiting the farms on their website and posts. Also, as Song pointed out (*ibid.*), foreigners are usually sensitive to food security issues and, more importantly, are more likely to have enough money to buy the products regularly. By scaling their project towards foreign customers, Farming is constructing a certain image and discursive history of themselves, and they thereby also *interested* and *enrolled* new Chinese customers.

Mobilization

According to Latour (2005), even when actors are *enrolled*, a network may stabilize itself for a certain period but it will likely require a certain level of solidity to reach long-term sustainability. By reaching out to a diversity of actants in both rural and urban areas, Farming publicizes its work and stabilizes its position in the longer term. Yet, this does not mean that, sooner or later, *enrolled* actants may not refuse to continue playing the "role" that was appointed to them. Because Song had established close relations with international actors and foreign friends in Shanghai, it became difficult for his organization to cut off these relations. Yet because of the Charity and Overseas NGO Management laws, collaborating with foreign partners and getting funds from them turned out to be a challenge. The desire to establish a social enterprise is an effort to diminish the organization's dependency on big retailers, but also to bypass legal restrictions. Besides, similar organic farming enterprises were booming all around China. Maintaining a certain flexibility was crucial to fight competition from fast developing urban and peri-urban farms, such as the Sunqiao Urban Agricultural District.[21]

In recent years, Farming has continuously adapted, stabilized, and innovated to respond to pressures and rapid changes. By presenting Farming as a social enterprise and building links with important international networks, Farming is more likely to survive because it can continue to rely on its biggest supporters. Recently, Farming's sister organization, based in Hong Kong, developed new projects that include eco-villages and eco-educational tourism. A focus on tourism is a new strategy to further enhance the relationships between urban and rural communities through direct visits. Visitors can experience the rural lifestyle (extending the relationship they already had via their smartphones) by visiting the Hunan mountains and seeing the farmers' "bucolic" lifestyles. These activities fortify Farming's network and distinguishes the organization from its competitors, and also uncovers new possibilities for rural areas to further empower themselves by opening hotels, selling crafts, or becoming tourist guides.

Overall, Farming solved its hurdles and limitations by constantly engaging in new and innovative solutions, and many other organizations I encountered in the field were developing similar strategies. They engaged in processes of *translation* to persuade other entities to adhere to their *problematization*. ANT illuminates the process by which organizations link social actors and technical entities through a diverse range of visible and invisible networks. Developing this type of socio-technical network is important in the Chinese context because it helps show how entities construct agency and (re)shape their socio-technical network if or when necessary. In China, observers often interpret social actors as entities surviving at the mercy of the state. One important lesson from organization Farming's project is that SGOs engage in "tactics" to "restabilize" their networks and resist top-down destabilizing pressures by allying themselves with other actants—human or non-human. They can thus create new relationships within Chinese civil society such as collective initiatives. Yet the marketization of SGOs leads to new limitations that reinforce the CCP's "green" consensus as highlighted above.

The limits of marketization

Despite the external controls faced by SGOs as they develop, the case study I present here proves that it is possible to embrace market strategies to side-step constraining laws. Yet, as I develop below, organization Farming's actions stay limited to social change and do not challenge the party-state's status quo. The Chinese government itself recognized a growing need to support rural communities and incentivize farmers to return to the countryside. The Shanghai Master Plan (2017–2035) issued in 2018 (discussed in more detail in Chapter 6) reflects the state's concern regarding overpopulated cities. Recent policies prove a desire to modernize the farming sector and boost rural incomes and living standards. The state's five-year plan for a rural vitalization strategy encapsulates the anxieties of the authorities on the urban–rural divide.[22] Interviewees working on various environmental issues highlighted the fact that it is difficult and competitive to get governmental funds. For those working on issues that are not at the forefront of the state agenda, engaging in multi-scalar strategies might be a good strategy as this may enable them to share their interests with new actors while diversifying their resources, as the case of Farming shows. But becoming independent from the government does not mean that these organizations are challenging the system or giving form to new, politicized forms of conduct. Indeed, in the section below, I argue the exact contrary by exploring the work other organizations and using Foucault to expand my analysis.

If we return to Foucault's understanding of the government as a "conduct of conduct(s)", we can see how this applies to our context, as by adopting market strategies, Farming redirected its action towards depoliticized forms of conduct. Indeed, by marketizing itself, in a way Farming ended up actually responding to Chinese leaders' anxieties. They not only become more independent from state resources, but they also reinforced the goal of rural vitalization by bringing images of modern and entrepreneurial farmers to the city sphere. To further develop this argument, I will now focus on the case of Turtle, which was briefly presented above. Turtle is a non-profit, volunteer-led organization created in 2009 in Shanghai. Registered as a for-profit company in Hong Kong until 2018, Turtle focused on informing ordinary people about sustainable models of growth and consumption. The organization has been advocating, for instance, for the sustainable management of freshwater resources. Their events included discussion of the limits of several projects, such as China's South-to-North Water Diversion project, which involves drawing water from southern rivers and supplying it to the dry north. Turtle organized events such as film screenings and discussions with experts, followed by brainstorming conversations with the public, creating a vibrant space for discussion among a diverse set of actors ranging from expats, Chinese entrepreneurs, and students to activists.

From 2016 to 2018, Turtle's events were increasingly organized in Chinese. Slowly over time, the leadership team—initially all foreigners—shifted to become half Chinese and half foreigners. After taking part in several of their activities, I came to realize why: Before registering in China, Turtle could not employ Chinese citizens. Some Chinese members attended, mostly students, but they worked solely as volunteers. After registering in the Mainland, however, Turtle changed its mode of operation. Its clients now include big companies such as Google, international banks, and big retail companies. Changing from a non-profit model focused on environmental advocacy towards a for-profit business also led to the complete restructuring of the organization's philosophy. Initially unstructured and advocating for change, the organization now focuses on doing "whatever can be done", setting aside questions such as "what is not being done?", or on "what is being done wrong?". The organization's discourses also became increasingly focused on individuals' responsibilities as opposed to those of states or companies.

Other organizations I observed (e.g., Goat, Mouse, and Koala) followed similar trajectories. Once surviving in a grey zone, when they adopted for-profit mechanisms this shifted their primary goal to be the "effectiveness" of their solutions, while advocacy and ideological thoughts fell by the wayside. In the case of Turtle, the more marketing they did, the more they focused on waste solutions (a key priority for the local government). Essentially, social entrepreneurs translate issues into a business model that turns social issues into market gaps. The social issues are then resolved by market strategy. In a recent book that explores the lifeworld of Contemporary social entrepreneurs in London and Milan, Carolina Bandinelli (2020) stresses that "social enterprises" epitomize the ethical struggle of our times. Bandinelli asserts that social entrepreneurs produce—and are produced—by post-political conditions because they reduce politics to the administration of things without engaging with the roots of a problem. This is exactly what I observed in Shanghai.

The more organizations engaged in marketing, the more their goals and ways of achieving them centered on sustainable economic and social development (*jingji shehui ke chixu fazhan* 经济社会可持续发展, in Chinese). Again, this closely follows President Xi's approach to "ecological civilization"; these are not dissenting voices nor do they create alternatives because their actions align with the main narrative of the party and the state.[23] Some even praise the "draconian" measures taken by the CCP to respond to issues such as waste management or fishing restrictions in the Yangtze River. Yet little to no attention is directed towards the destructive practices of Chinese fishing vessels in Africa or Latin America. Events increasingly slip into depoliticized arenas, favoring discourses that emphasize the "innovative" and "entrepreneurial" image of Shanghai which Chinese authorities try so hard to publicize—I will further assess this point in Chapter 6 by taking Shanghai's master plan as a case study.

Let the market decide

Establishing a social enterprise and relying on market strategies is, thus, not a panacea to all China's civil society problems. As many practitioners have emphasized, establishing a social enterprise creates new challenges and constraints. From a lack of sustainable and effective funding mechanisms, difficulties in recruiting talent, limited resources, and services, to insufficient business skills, it is a fact that few of the observed social enterprises were able to successfully scale their projects. Because there is no formal recognition of their status, there are no pipeline development opportunities to help them develop. As happened with NGOs for nearly 30 years, social enterprises are trapped in a "grey" zone. By creating the means for SGOs to (re)embrace a "hybrid" identity, social enterprises were purposefully used to blur the public-private line.[24] This hybridity is not fixed and may develop in response to changing pressures and demands. Yet although the elasticity of a market identity seems to give some SGOs alternative paths for development, embracing neoliberal rationalities considerably weakens their role as agents of change.

As observed in the field, those organizations working on sensitive issues were the ones that struggled the most because it was difficult for them to make a profit when working on discrimination or poverty. Even organizations working on non-sensitive issues such as waste were not immune to hurdles, as the case of Goat illustrates. In 2009, Goat (see Appendix C) registered as a non-profit organization in Hong Kong under the Shanghai Overseas Chinese Foundation. At some point, the project was in decline, so the foundation wanted to close it. The founders, still seeing potential in the organization, decided for the project to become a social enterprise (while maintaining legal non-profit identity). The aim was to offer a professional and qualitative program that private companies could sponsor. After a successful launch in 2012, they opened a branch organization (or operational company, as they put it) in Mainland China to support the project's development (as a for-profit company). I'll call it Mini Goat. Yet, both organizations had the same office in Shanghai. So, it was common for people working at Mini Goat to use Goat's name (more prestigious in Shanghai). In 2017, Goat had to make a public declaration regarding the misuse of the organization's logo on several collaborative projects they had started with Turtle (which was still registered in Hong Kong at the time). At the heart of the "scandal" were two foreign employees, who (officially) worked in their branch organization. The persons in question were accused of misusing the name of Goat for fraudulent purposes. I had met one of these people at the headquarters of Goat where they had been working for more than a year. The communiqué stated the following:[25]

> The mention of the "[XX] project launch" by [Mr./Ms. A] on their personal profile on LinkedIn and any other web communications, during

> their tenure as a General Manager of [Mini Goat] (August 2015–April 2017) is fraudulent. They were never appointed General Manager of [Goat], although, they frequently misused this signature in emails and external communication for fraudulent purposes. In their former capacity of [Mini Goat], they also diverted our supporters' financial contributions to another bank account than that of their employer.

Goat was established in 2012 in Hong Kong, but its branch organization was registered in China. Both organizations have the same primary address, use the same website and the same email account. While Goat was officially a non-profit organization, which means that it needs to follow certain rules to be able to develop its activities in the Mainland, Mini Goat, as a private enterprise, enjoys more leeway. Due to the fact that Goat, and not Mini Goat, was collaborating with a foreign company without duly reporting it to the Chinese authorities (as specified in the guidelines of the Charity Law), the organization was officially breaking the law.

This strategy, of using double identities, is like the one established by Farming as mentioned above. Working with vague identities makes it difficult even for people working for the organization. The different parties never objectively explained why they had to make the public declaration cited above, but this directly resulted from a misuse of their legal identity fabrics. Mr./Ms. A has since then left the organization.

Another issue that could affect the sector is the nature of the Chinese government, as future government legislation could seriously influence growth. This happened recently with NGOs as well as social welfare enterprises. The China Volunteer Service Law (*zhiyuan fuwu tiaoli* 志愿服务条例),[26] for example, which took effect in 2017, severely affected the work of NGOs and Corporate Social Responsibility programming in China, and also served to remind us that the Party will maintain its grip on the sector.

Moreover, by not being formally recognized, the money such organizations make on sales or services, and the way they use their money isn't supervised by any official entities. Everything is based on trust. A scandal of money mismanagement, as previously happened with the Red Cross,[27] could seriously tarnish an organization's reputation, and can also create frictions and misunderstanding among Chinese citizens as big, profitable organizations increasingly engage in social causes to gain legitimacy among customers.

Likewise, as stressed above, it is very difficult for organizations focusing on advocacy to develop a viable business model, even if they use a social enterprise identity. Representatives from the organizations discussed here (except Wolf and Whale) explained that although they do not meet the legal requirements to formally register as an NGO—because they focus on sensitive issues and maintain a close relationship with foreign actors (e.g., in the case of organizations such as Tiger, Panda, Bear, Eagle); or because they cannot respond to the Overseas NGO Law requirements (e.g., Snake, Turtle,

Mouse, Goat)—they are nevertheless obliged to take this path to survive, even though their chances of surviving were extremely small.

For some organizations working on sensitive issues (labor rights, LGBT, etc.), adopting an entrepreneurial approach is very difficult and, even when the organization develops a viable business model, the risk is that whatever issue they are working on will become depoliticized. Meanwhile, as two interviewees pointed out (NPI Marketing and Communication Manager, October 2016; Communication Manager SynTao, April 2017), Shanghai's local government is pushing organizations to embrace the path of social marketization and no longer rely on government funds.

Regarding the last point, it is important to highlight that SGOs and the state are both aware of one another. As the founder of Turtle explained, the local government tolerates and sometimes supports their activities (Interview April 2016):

> As long as you don't show the government in a negative light, as long as you don't do controversial stuff it's perfectly fine. I think bigger organizations like Greenpeace, for them there is a little more at stake. They are always in the government agenda because they are really big organizations… and the things they do get noticed by a lot of people. So, naturally, the government is going to be much more focused on them. But small organizations like ours you know […] Thus, so far, we didn't have any issues.

Indeed, most of the foreign SGOs observed here were small organizations, sometimes comprising one or two full-time employees and relying heavily on volunteers and interns. Given their size, the majority did not see themselves as threatening to the party-state and felt safe regarding to their activities. Two of the organizations were even asked by local leaders to collaborate with them, and the only reason for them not to do was their size and lack of professional capacity. Yet, as highlighted in this chapter, SGOs never know when they will cross the "red line". Besides, as Kroeber (2016) argues, economic growth has become crucial for the party to strengthen its grip on power as people feel more secure sticking with a system they know rather than trying another. For instance, the organization Mouse that has been working on zero waste solutions since 2017 felt that it needed to avoid entering confrontational discourses with the capitalist and consumerist path taken by China (I explained in Chapter 2 how many activists were detained because they had questioned the logic of the CCP's economic development strategy). They preferred to focus on the circular economy to avoid sensitive issues such as degrowth or downscaling of production and consumption.

Roughly, one can conclude that for now, social enterprises are not a threat to the party-state. I observed two tendencies related to the fact that these organizations have to compete within the laws of the market. First, they focus on issues that touch upon sensitive or advocacy issues and so they are not marketable and remain weak; or, second they develop successful market strategies but then increasingly end up focusing on depoliticized topics.

Conclusion

Given the current regulatory environment under Xi Jinping's government, it is important to understand how Chinese society is adapting to new pressures. The present chapter had two major goals: First, to show that, even in a controlled environment, SGOs develop at the intersection of different sectors and institutional logics by constantly redefining and renegotiating their boundaries. For Chinese SGOs, whether or not they are working on sensitive issues, this enables them to develop with more autonomy by relying less on the state for resources. For foreign SGOs registered as companies or unable to fulfil new legal requirements, social marketization opens new prospects for action.

Second, this chapter has demonstrated that ANT is useful as a methodological tool to analyze current bottom-up environmental initiatives in authoritarian settings. We have seen, for instance, that entities and their attributes are an effect of an organization's relations with other entities, rather than inherent properties (Law, 1999; Rutland and Aylett, 2008). This analysis brought to light the roles of discourses, technologies and infrastructures as implements of societal empowerment, reinforcing the stereotypical view that members of Chinese civil society are agentic subjects with little capacity for individual action. The organization Farmer later decision to scale and to develop agency by engaging with the market was, on the one hand, a direct response to several limitations the organization had to face and, on the other, a way to take advantage of the opportunities that a social enterprise entity could open to them at that time.

Still, it would be wrong to claim that independence and governmental pressures are the only factors that prompt organizations to adopt a for-profit model. The marketization of SGOs emerged and spread around the world in the past three decades. Social entrepreneurship and the acquisition of government contracts to purchase services are two salient indicators of the new tendency of social marketization at a global level. Even though regulations at a central state level are reshaping SGOs' strategies, we should not neglect global narratives in our understanding of why and how civil society is growing in certain ways. Even in a regime as authoritarian as that of China, global norms impact the state, the economy, and society at different levels. We should therefore consider both fragmentation and globalization when studying SGOs and bottom-up networks in China because they create new partnerships of partnership and community development such as new public "responsibilities" and priorities (Leutert 2018, 190).

As a new concept, this new hybrid corporation model deserves acute attention, especially in a cosmopolitan city like Shanghai. In the last few years, an emerging network composed of incubators, universities, co-working spaces, consulting companies, and research centers have been pushing their development. New organizations (e.g., Impact Hub Shanghai, established in late 2016 in Shanghai) are reinforcing collaborations and exchanges between Chinese and foreign SGOs. These new hybrid organizations, while not offering a full alternative to the formal social welfare system, act as a tool to create fairer, accountable, and more sustainable cities by advancing new solutions to problems left unanswered by the party and the state.

Yet because of a lack of access to funds from non-private actors and, thus, the need to follow market rules, they risk falling into depoliticized forms of participation and, thus, reinforcing the "green" consensus advanced by the CCP. Chapters 3 and 4 particularly highlighted how Shanghai has become a pioneer in shifting the role of the government in public services and influencing the development of SGOs to fill the gap in social services. I particularly showed that, as one of the fastest-developing cities in the world, Shanghai must find diverse welfare provision practices while continuing to situate itself in complex, central-local relationships and also within the heterogeneous local, political, and socio-economic context. So, the growth of social enterprises can end up enhancing authoritarianism. A social marketization strategy can create ambivalence and apolitical narratives whereby everything and nothing is political at the same time. By associating their goals with an increasingly pluralistic, multicultural, complex, and transnational identity, it is more difficult for SGOs to focus on highly political issues and influence policymaking. Moreover, the party-state's growing involvement with strategic partnerships, symbolic recognition (e.g., awards or prizes), human capital formation (e.g., training), or regulations could lead to patterns of co-optation and (again) strong reliance on government support.

In conclusion, although it can be tempting to consider the marketization of SGOs as proof of their emancipation, the reality on the ground is that organizations risk falling into ambivalent and apolitical narratives that reaffirm the goals of the party: thereby nurturing a model of "sustainable urbanism", as further explored in the following chapter. Hence, it is essential to conduct studies that further explore how the growing marketization of the sector—and the inherently apolitical strategy—may impact the political/advocacy goals of SGOs in the near future.

Notes

1 According to de Certeau, people are not passive and engage in "tactics" to resist the material and symbolic productions of the power or the dominant system. I will engage more deeply with de Certeau's work in Chapter 6.

2 Live streams of DBS Foundation and Social Enterprise Research Centre Report Release: Report on Social Enterprises in Greater China [*da zhonghua qu shehui qiye diaoyan baogao* 大中华区社会企业调研报告]. (accessed 21 February 2017).

3 See research by EMES, available at https://emes.net/.

4 The Peking University China Social Work Research journal translated the first academic journal "The Social Enterprise"; Sino-Symposium on Social Enterprise and NPO organised in Beijing.

5 See British Council: "Social Enterprise Programme", available at https://www.britishcouncil.cn/en/programmes/society/social-enterprise-programme (accessed 27 March 2017).

6 The Foundation for Youth Social Entrepreneurship (FYSE) report provides an insight into the current state, challenges, and opportunities of social entrepreneurship in China.

7 This research was jointly researched, written, and published by Shanghai University of Finance & Economics Social Enterprise Research Centre, Peking University Centre for Civil Society Studies, the 21st Century Social Innovation Research Centre, and the University of Pennsylvania School of Social Policy and Practice.

8 Data retrieved from Ding Li's presentation at the Green Drinks Chinese Forum titled "The State of Philanthropy in China [*zhongguo cishan xianzhuang* 中国慈善现状]", 22 December 2016, Shanghai. Ding Li argued that social enterprises should "have their own charter and council, rather than being part of the government, charitable organizations or other agencies [*you ziji duli de zhangcheng he lishi hui, er bushi zhengfu, cishan zuzhi huozhe ta jigou de yibufen* 有自己独立的章程和理事会，而不是政府，慈善组织或者他机构的一部分]".

9 Approximately one in every four foreigners in China lives in Shanghai. According to the Shanghai Municipal Statistics Bureau, there were 171,874 foreigners officially registered in Shanghai in 2014.

10 Su Zhou. 2016. Shanghai tops the list of most attractive cities for expats. *China Daily*. https://www.chinadaily.com.cn/china/2016-04/16/content_24599257.htm (accessed 16 February 2017).

11 Daniel Ren (2017) Tesla's US$5 billion Gigafactory helps Shanghai record increase in foreign direct investment for 2018. *South China Morning Post*. https://www.scmp.com/business/companies/article/2183222/teslas-us5-billion-gigafactory-helps-shanghai-record-increase (accessed 11 November 2019).

12 (2019) China Cuts Taxes in Shanghai Free Trade Zone to Lure Investment. *Bloomberg*. https://www.bloomberg.com/news/articles/2019-08-30/chinacuts-taxes-in-shanghai-free-trade-zone-to-lure-investment (accessed 10 November 2019).

13 See Global Legal Monitor (2019) "China: Foreign Investment Law Passed", available at https://www.loc.gov/law/foreign-news/article/china-foreign-investment-law-passed/ (accessed 30 September 2019).

14 Refer to Zoey Zhang (2019) "Shanghai Releases Implementation Regulation for Foreign Investment Law", China Briefing, available at https://www.china-briefing.com/news/shanghai-releases-implementation-regulation-foreign-investment-law/ (accessed 30 September 2019).

15 Source: Hogan Lovells, an American-British law firm co-headquartered in London and Washington, DC. Refer to "China's New Law on Foreign NGOs: will my organization need to pay (more) taxes in China?", available at https://www.hoganlovells.com/~/media/hogan-lovells/pdf/news/2016/chinas-new-law-on-foreign-ngos-tax-implicationsshalib01.pdf?la=en (accessed 14 May 2019).

16 Presentation by Ding Li titled "Shanghai's Social Enterprise Landscape: Reflections on the Past and Present" at the Fresh Start Rotary Club of Shanghai meeting, 26 May 2017, Shanghai.

17 For more information refer to the Permaculture Project website: https://www.permacultureproject.com/ (accessed 12 September 2019).

18 Song's presentation at a public conference, August 2016.

19 Children who remain in rural regions of China (normally with their grandparents) while their parents leave to find work in urban areas.

20 Slow Food is a global, grassroots organization, founded in 1989 to prevent the disappearance of local food cultures and traditions, counteract the rise of fast life, and combat people's dwindling interest in the food they eat, where it comes from and how our food choices affect the world around us. Refer to the Slow Food webpage available at http://www.slowfood.com.

21 For more information about the project see http://www.sasaki.com/project/417/sunqiao-urban-agricultural-district/ (accessed 12 December 2019).

22 See http://english.gov.cn/policies/latest_releases/2018/09/26/content_281476319507892.htm (accessed 12 October 2018).
23 See the Ecological Civilization of China website available here: http://www.cecrpa.org.cn.
24 The book *China's Youth Cultures and Collective Spaces*, edited by Frangville and Gaffric (2019) demonstrates through several case studies that youth cultures navigate between the formal and the informal.
25 The communiqué was taken from the organization's official website. Due to the sensitivity of the issue, I have erased all information that could allow the organization to be easily recognised. The modified parts are in square brackets.
26 (22 August 2017) "志愿服务条例 [Voluntary Service Regulations]". *State Council of the People's Republic of* China, available at http://www.gov.cn/zhengce/content/2017-09/06/content_5223028.htm. (accessed 12 January 2018).
27 See Celia Hatton (22 April 2013), "China's Red Cross fights to win back trust", *BBC News*, Beijing, available at https://www.bbc.co.uk/news/world-asia-china-22244339 (accessed 23 April 2018).

References

Bandinelli, Carolina. 2020. *Social Entrepreneurship and Neoliberalism: Making Money While Doing Good.* Rowman & Littlefield International.

Boelens, Rutgerd, Jaime Hoogesteger, Erik Swyngedouw, Jeroen Vos, and Philippus Wester. 2016. "Hydrosocial Territories: A Political Ecology Perspective". *Water International* 41 (1): 1–14.

Carrillo, Beatriz, and Jane Duckett. 2011. *China's Changing Welfare Mix: Local Perspectives.* Routledge.

Dart, Raymond. 2004. "The Legitimacy of Social Enterprise". *Nonprofit Management and Leadership* 14 (4): 411–424.

de Certeau, Michel. 1984. "Walking in the City". In *The Practice of Everyday Life*, 91–110. University of California Press.

Ding, Dang, Pingyang Liu, and Neil Ravenscroft. 2018. "The New Urban Agricultural Geography of Shanghai". *Geoforum* 90: 74–83.

Frangville, Vanessa, and Gwennaël Gaffric. 2019. *China's Youth Cultures and Collective Spaces.* Routledge.

FYSE. 2012. *China Social Enterprise Report.* http://www.chinadevelopmentbrief.cn/wp-content/uploads/2014/10/China-Social-Enterprise-Report-2012.pdf.

Hogan Lovells. 2016. *China's New Law on Foreign NGOs: Does It Apply to You, and If so, What Do You Need to Know*? https://www.hoganlovells.com/~/media/hogan-lovells/pdf/news/2016/chinas_new_law_on_foreign_ngos-does-it-apply-to-you.pdf?la=en.

Hsu, Carolyn L. 2017. "Social Entrepreneurship and Citizenship in China". *The Asia-Pacific Journal* 15 (3).

Huang, Zuhui. 2013. "Farmer Cooperatives in China: Development and Diversification". In *The Oxford Companion to the Economics of China.* Oxford University Press.

Huybrechts, Benjamin, Jacques Defourny, M. Nyssens, T. Bauwens, O. Brolis, P. De Cuyper, F. Degavre, et al. 2016. "Social Enterprise in Belgium: A Diversity of Roots, Models and Fields". http://orbi.ulg.ac.be/handle/2268/183267.

Huysentruyt, Marieke. 2016. *Country Report China. Brussels and Shanghai.* Seforïs.

Keping, Yu. 2011. "Civil Society in China: Concepts, Classification and Institutional Environment". In *State and Civil Society: The Chinese Perspective*, edited by Deng Zhenglai, 88–89. World Scientific.

Kim, S. Y., and Jacques Defourny. 2011. "Emerging Models of Social Enterprise in Eastern Asia: A Crosscountry Analysis". *Social Enterprise Journal* 7 (1): 86–111.

Kroeber, Arthur R. 2016. *China's Economy: What Everyone Needs to Know*. Oxford University Press.

Latour, Bruno. 1996. "On Actor-Network Theory. A Few Clarifications, Plus More Than a Few Complications". *Soziale Welt* 47 (4): 369–381.

Latour, Bruno. 2004. *Politics of Nature. How to Bring the Sciences into Democracy*. Harvard University Press.

Latour, Bruno. 2005. *Reassembling the Social: An Introduction to Actor-Network Theory*. Oxford University Press.

Law, John. 1999. "After ANT: Complexity, Naming and Topology". *Sociological Review* 47 (1) (suppl.): 1–14.

Leggett, Angela. 2017. "Bringing Green Food to the Chinese Table: How Civil Society Actors Are Changing Consumer Culture in China". *Journal of Consumer Culture* 20 (1): 83–101.

Leutert, Wendy. 2018. "Firm Control: Governing the State-Owned Economy Under Xi Jinping". *China Perspectives* 2018: 27–36.

Li, Ding, and Anne Merkle. 2017. "Talent: The Key to Developing the Social Sector in China". *Stanford Social Innovation Review*. https://ssir.org/articles/entry/talent_the_key_to_developing_the_social_sector_in_china#.

Oberoi, Roopinder, Ian G. Cook, Jamie P. Halsall, Michael Snowden, and Pete Woodock. 2019. "Redefining Social Enterprise in the Global World: Study of China and India". *Social Responsibility Journal* 16 (7): 1001–1012.

Pallotta, Dan. 2010. *Uncharitable: How Restraints on Nonprofits Undermine Their Potential*. University Press of New England.

Poon, Daryl. 2011. *The Emergence and Development of Social Enterprise Sectors*. Working paper. https://repository.upenn.edu/sire/8.

Rutland, Ted, and Alex Aylett. 2008. "The Work of Policy: Actor Networks, Governmentality, and Local Action on Climate Change in Portland, Oregon". *Environment and Planning D: Society and Space* 26 (4): 627–646.

Scott, Stephanie, Zhengzhong Si, Theresa Schumilas, and Aijuan Chen. 2018. *Organic Food and Farming in China: Top-down and Bottom-up Ecological Initiatives*. Routledge.

Song, Yiching, Gubo Qi, Yanyan Zhang, and Ronnie Vernooy. 2014. "Farmer Cooperatives in China: Diverse Pathways to Sustainable Rural Development". *International Journal of Agricultural Sustainability* 12 (2): 95–108.

Swyngedouw, Erik, Nikolas C. Heynen, and Maria Kaika. 2014. "Urban Political Ecology. Great Promises, Deadlock … and New Beginnings?" *Documents d'Analisi Geografica* 60 (3): 459–481.

Tauber, Lilian. 2019. "Beyond Homogeneity: Redefining Social Entrepreneurship in Authoritarian Contexts". *Journal of Social Entrepreneurship* 12 (3): 1–19.

Wang, H., I. Alon, and C. Kimble. 2015. "Dialogue in the Dark: Shedding Light on the Development of Social Enterprises in China". *Global Business and Organizational Excellence* 34 (4): 60–69.

Wu, Jing. 2018. "How a Social Enterprise Balances Philanthropy and Commerce?—Perspective of Institutional Logic [Shehui qiye ruhe jiangu gongyi yu shangye——

jiyu zhidu luoji de fenxi 社会企业如何兼顾公益与商业——基于制度逻辑的分析]". *Social Science of Beijing* (10): 119–128.

Zhang, Jiawei, and Sherry Shi. 2017. "Social Enterprises in Greater China States (Da Zhonghua Qu Shehui Qiye Diaoyan Baogao 大中华区社会企业调研报告)". http://www.serc-china.org/services/overview/data.html.

Zhao, Meng. 2012. "The Social Enterprise Emerges in China". *Stanford Social Innovation Review* 10 (2): 30–35.

Zhao, Zhiyuan. 2014. *The State of Entrepreneurship in China: Seforïs Country Report*. https://static1.squarespace.com/static/56d2eebbb654f9329ddbd20e/t/5773e650f5e231dfe731c18e/1467213393773/Country_Report_China.pdf.

Zhou, Weiyan, Xiaobin Zhu, Tianxue Qiu, Ruijun Yuan, Jingya Chen, and Tongkui Chen. 2013. *China Social Enterprise and Impact Investment Report*. https://www.scribd.com/document/165006026/2013-China-Social-Enteprise-and-Impact-Investment-Report.

6 Urban sustainability as consensual practice

Introduction

Observers commonly assume that the heart of the CCP's approach to state-led environmental policy centers on coercive and Draconian mechanisms, such as target-setting. As I showed in the previous chapters, however, *environmental authoritarianism* is infused with forms and practices of "participation" and "inclusion", or what I refer to as cooperative governance processes (see Chapter 3). Based on my empirical observations in Shanghai during my fieldwork with environmental SGOs, I argue that environmentalism is giving new opportunities to Chinese leaders to co-opt and mobilize a variety of actors at the local level with the aim of increasing authoritarian control. "Green" discourses function as a pillar to place consensus at the heart of state–society relations and enforce citizen engagement that benefit the regime's goal of reaching an "ecological civilization" (as shown in Chapter 4 with the case of waste management) to, ultimately, maintain the CCP in power. Cities have come to play a major role in performing and displaying this consensus model of sustainability, and it is thus useful to consider more deeply to what extent cities epitomize the country's recent developments in environmental governance (Pow 2018).

In this chapter, I will continue to focus on Shanghai as an avenue to further explore the underbelly of *environmental authoritarianism*, associated with China's non-democratic approach to public policy (Beeson 2016). Here I want to stress the importance of cities in observing how China's purpose of sustainable development goals are attained through consensual modes of policy making articulated around public, private, and non-state partnerships operating in a frame of agreed/tolerated objectives (working on issues such as waste management, environmental education, or city greening). I will focus here in particular on the organizations' political and practical resolution, and how they have become crucial for the state to claim legitimacy and limit dissent. As Wells and Lamb depicted in their work on Myanmar (2021)—a country increasingly recognized for its strong authoritarian regime since the military coup in 2021—my observations show that decision-makers dilute and defuse disagreements through the construction of a utopian vision of a

DOI: 10.4324/9781003231325-6

modern "Excellent Global City" to which environmental SGOs and citizens are deemed to contribute. My aim is to put into perspective the theoretical claims made in Chapter 2, and also show how urban imaginaries of the "sustainable" city function as the perfect backdrop for *environmental authoritarianism*.

There is no intention here to universalize the findings; rather I zoom in on Shanghai's 2035 master plan to assess sustainability as a process of strategic hegemonic sedimentation. Scholars consider Shanghai 2035 as epitomizing the development of the city's iconic status as a model for other Chinese and foreign cities to follow in the future (Qiyu 2019; Den Hartog 2021). Because the greening of cities—or eco-city projects—serve as new planning paradigms to bring different societal aspects of urban development into a consensual political strategy (Mössner 2016), they act as good lens to scrutinize the new approach of Chinese leaders, moving from coercive mechanisms towards a consensus method. This incarnates the post-political condition identified in the previous chapters.

Although studies of depoliticization have mainly focused on assessing the logic of capitalism and the hegemony of market-liberalism in liberal democratic settings, scholarship increasingly recognizes such processes as essential to actors pursuing other political agendas (Rancière 1995b). While Lendvai-Bainton and Szelewa (2021) acknowledge and conceptualize the rise of authoritarian neoliberalism as a particular and geographically specific form of the post-political in Eastern Europe; Lawreniuk (2021) provides a grounded account of the operationalization of post-politics in processes of urban development and authoritarian neoliberal governance in Cambodia. Similarly, Turhana and Gündoğan (2017) note that hegemonic and techno managerial narratives are colonizing Turkey's green economy, which is leading to processes of post-politicization. This chapter is a follow-up to these preliminary studies in authoritarian, post-authoritarian and non-European contexts (Wells and Lamb 2021), claiming that there is no singular post-political condition.

While the rise of eco-cities in China has attracted considerable scholarly attention, especially in critically assessing their limits and contradictions (Pow and Neo 2013; Caprotti 2014a; Li and de Jong 2017), little attention has been paid to dissecting the "condition of post-politics" in a Chinese context. As explored by Rancière, Žižek, and Mouffe in Europe, or Harvey Neo with regard to Singapore (Neo 2021), I argue that Shanghai increasingly suffers from the "post-political" condition, as the urban landscape is increasingly "colonized or sutured by consensual techno-managerial policies" (Swyngedouw 2011). My interest here is the political dimension of urban sustainability and how such processes change the nature of civil society actors and individuals. Why and how cities have become vectors of depoliticization in contemporary China?

To answer this question, I will start by exploring how cities are increasingly being constructed to showcase China's "green" ambitions. In particular, I take Shanghai 2035, the city released State Council-approved master plan for the years 2017–2035, as a case study to support my claims. I will highlight three points: First, the inconsistencies behind policymakers' claim to involve the public in governance and decision-making in urbanization processes; second, how such narratives of the "sustainable" city reinforce the post-political condition; and third, despite a trend of depoliticization, I build on de Certeau's concept of "strategies" and "tactics" to claim that small, on-the-ground acts of resistance should be given more attention.

The "greening" of Chinese cities

Given its authoritarian context, China's urbanization has been dictated by the political incentives the central government has been (mindfully) providing (Logan 2018),[1] "to position themselves (cities) advantageously on the global scene" (Rosol et al. 2017, 1710). Although competition has been directed towards boosting economic growth (at any cost)—bringing China near to environmental collapse or, as Richard Smith (2020) termed it, "eco-suicide"—more recently, the government has set greener requirements (see Chapter 3). From the beginning of the 2010s, the Chinese government signaled a more holistic approach to the environment (Termine and Brombal 2018) alongside a commitment to pursue a new phase of urbanization to tackle years of environmental mismanagement. The release of the National Urbanization Plan in 2014 (2014–2020), for instance, prompted new methods of urbanization, including environmentally friendly approaches to urbanism (Cheshmehzangi 2016). Yet, as I explain below, the promotion of environmental sustainability is also being led by initiatives that aim to optimize urban expansion, such as controlling the excessive growth of metropolises like Shanghai (Chu 2020).

Although the party-state's concern for urban sustainability is not new (Verdini 2015), recent years saw a proliferation of policies and practices for sustainable urban development, such as the adoption of concepts like "ecological civilization" (Gare 2017; Termine and Brombal 2018). Within this strategy, cities act as branding tools to advance China's "green" ambitions (de Jong et al. 2016). A good example of this shift and the growing importance being given to urban areas is the adoption of urban sustainability indicator frameworks, such as the China Urban Sustainability Index.[2] Performance evaluations range from including environmental quantitative targets to digital monitoring tools under the increasing authority of the Ministry of Ecology and Environment. Such new, vertical layers of governance are in line with the coercive measures explored in Chapter 2 and have sparked the attention of scholars, with some arguing that this is leading to the development of a regime of green urbanism (Hoffman 2011; Ren 2012; Miao and Lang 2015; Brehm and Svensson 2017).

As Pow and Neo (2015) pointed out, ecology has become another field to be tamed, managed, and mobilized to follow national economic agendas and local entrepreneurial goals. The processes by which China is directing its narrative construction towards a more environmentally friendly future is well captured by the way "eco-cities" are being fashioned and by all their environmental imaginaries and discursive formations. As Caprotti (2014b) showed with the case of Tianjin, "eco-cities" are ecological anchors for the reproduction of neoliberal practices and, thus, reflect the multiple frictions and clashes in contemporary Chinese society. Recent literature reviewing urban management and planning in China particularly emphasizes the prevalence of ideologies of pro-growth entrepreneurial development and highly "technocratic" approaches to urbanization (Pow and Neo 2015; de Jong et al. 2016). Similarly to Caprotti's analysis of Tianjin, Zhang and Wu's (2021) study on Taihu New Town stresses the multiple contradictory political, economic, and legal interests involved in such eco-flagship projects.

While a need to standardize sustainable urban practices becomes a necessary precondition to address a range of issues, from enhancing quality of life to sustaining capital accumulation, there is also a strategy of distancing China's urban experience from Western rhetoric (Wang-Kaeding 2021). Because China's urbanization has been well received at an international level (leaving aside its episodes of smog and food scandals), Chinese leaders want successful urbanization initiatives to bear the imprint of the power and influence of the one-party system (Pow 2012, 49). This follows, as historian Zhang Lifan claims, the "partyfication" (*dang hua* 党化) of everything.[3] Likewise, Xu (2018) contends that, although "Chinese exceptionalism" resisted the attacks of universal values, more recently, a "China model" has been put forward to seize the discursive power of global civilization.[4]

The promotional narrativization of "sustainable" urban imaginaries

The use of the environment as a political tool in China is not new. It dates to Mao's leadership, when *verdurization* (*lühua* 绿化) campaigns were used "as part of its (Chinese State) project of building a socialist state and as a way of exercising ideological control, particularly in cities" (Lu 2017).[5] Despite this recognition of the role that greening campaigns have played in post-reform China, the politicization of environmental governance at an urban level have sparked little engagement in the literature. Focus has rather been on—amongst other things—the politicization of history (Wei 2008), stock markets (Li and Zhou 2016), Confucianism (S. Wu 2015) or, more recently, global warming (Eberhardt 2015; Shi and Wu 2018), and entrepreneurialism (see Fulong Wu's ERC project "Rethinking China's Model of Urban Governance"). It is notable, however, that despite wide "green" state discourses at a national level (and more recently, at an international level), beyond a few initial analyses stressing the discourse of sustainability as economically motivated, and "environmental imagineering" (Shiuh-Shen 2013; Caprotti 2014a cited in Wu 2020), the promotional narrativization of "green" urbanism remains underdeveloped. In contrast, a growing number of studies in Europe and North America show that the concept gives important clues to uncover the political interests, desires and anxieties circumventing urban construction.

China's rapid rise as the world's second-largest economy has resulted in an unprecedented redevelopment of urban areas (Wu and Zhang 2008). Drastic and violent urbanization plans are particularly visible in Shanghai (Wu 2016), one of the major economic and population centers of China. Tons of explosives have been used to demolish large parts of the city since the 1990s (Ping 2019), disrupting local identities and resulting in high financial, social, and human costs (Thomsen et al. 2011). This large-scale urban renewal not only transformed the appearance of the city's environment but altered people's lives and livelihoods as well as the way urban space is inhabited (Chu 2015; Wu 2016). This "urban modernity" or "civilized way of life", as Gibert and Segard (2015, 8) put it, "are used as a persuasive ideology and linked to an official discourse about order, reminding the idea of 'harmonious society' (*hexie shehui* 和谐社会), introduced in China by Hu Jintao".

For a long time, policymakers have operationalized cities through authoritarian processes. This is not a particularity of China. Be it in democratic or authoritarian political settings, cities "that work" have been shaped through displacement and segregation policies, limitation of movement, unequal flows of capital, or the destruction of public space (Fowler 1992). But even though ideal imaginaries of cities are nothing new, the urgency of sustainability is creating new opportunities for urban planners to enact new ideological visions of the city, for which there is no alternative (Swyngedouw 2010; Wilson and Swyngedouw 2014). As Ahlers and Shen (2018, 305) stress, this can lead to openings for the increased use and greater acceptance of top-down decision-making as the over-riding priority of sustainability is not only environmentally motivated but also helps maintain state power. Most times, this means mobilizing the "self-responsabilization" of individuals and consent from lower-level authorities and residents, creating openings for those in power to pursue their agenda. Thus, consensus is curated rather than enforced. It follows Foucault's concept of *governmentality*, which understands "power" to be about the "management of possibilities" and the ability to "structure the (possible) actions of others" rather than recourse to coercion (Foucault 2003 cited in McKee 2009).

The use of top-down and coercive methods—or, as Kostka and Zhang (2018) put it, a "tightening grip" on environmental governance—has already been extensively explored in the literature. China's emerging use of ad hoc campaigns, new technologies (e.g., big data, smart cities), central stringent inspections (*huanbao ducha* 环保督查), or party discipline in particular are recognized as central tools for the party and the state to address the country's environmental crisis (Kostka and Nahm 2017; Ma 2017; Jia and Chen 2019; Shen and Jiang 2021). Meanwhile, state–society relations in cities are increasingly shaped by cooperative governance mechanisms, as I explored in the previous chapters. Yet, a missing point of my argument, which I touched upon only briefly in Chapters 4 and 5, is the realm of routine practices, as operationalized through the political and aesthetic aspects of the administrative ordering of everyday life as the basis for asserting new forms of "green" citizenship and power (de Certeau 1984; Lefebvre 2000). The following section explores the question of "eco-city" imaginaries, and considers criticism of urban ideology related to current urbanization processes.

In this section I address the relationship between the production, and imagination, of urban space and politics. The more time I spend in the field experiencing the routine spaces of the city, the more I realized the importance of uncovering these depoliticization practices, not through coercive and punitive mechanisms, but rather through the proliferation of governance actors and the increasing call for "self-responsabilization". According to post-political scholars, such processes align with the post-political condition. Indeed, the promotion of increased participation in governance, as I observed when following the work of SGOs in Shanghai communities, did not question the existing status quo, but increasingly limited the space where it could be challenged. This sheds light on the shifting governmental logic and planning epistemologies that frame the concept of "ecological civilization" which links sustainability with the governing of people and, thus, reframes power dynamics.

There is widespread use of utopian images in urban development in China and we can see a certain parallel here with Swyngedouw's (2009, 2019; Swyngedouw and Wilson 2014) observation of the "disappearance of the political" in Europe. Yet, contrary to Europe where consensus is linked to neoliberalism and is upheld to keep the elite status quo, in China, consensus is used to uphold the leadership of the Communist Party of China, defined as the only realistic and practical pathway for building "China into a great modern socialist country that is prosperous, strong, democratic, culturally advanced, harmonious, and beautiful and realizing the Chinese Dream of national rejuvenation".[6]

Pow (2018) in particular has sketched out the emergence of an eco-aesthetic that selectively incorporates aesthetic principles into everyday governmental practices of China's urban planning. Again, this observation is not entirely new. By the 1970–1980s, state propaganda was being used to make people, both in the countryside and in the cities, adopt certain "morality" guidelines (e.g., keep surroundings clean, "beautify" the environment or protecting biodiversity).[7] Although the format has evolved in the digital age to reach as far as groups on WeChat, the narrative is not very different from that of the past.[8] As in the past, the general goal now is to define "good" behavior, and that includes making urban Chinese citizens recycle their waste but also non-environmental issues, such as publicly shaming those wearing pajamas in public—strolling on the street in sleepwear is common in China—or replacing Shanghai's Street food heritage with cleaner international chains that better fit the new, international, face of the city.[9]

So, as policymakers increasingly capitalize on sustainability to guide urban development (Pow 2018; Wang 2019), it becomes crucial to assess the trends and strategies lying behind such discourses to understand the risk posed by maximizing policing under the guise of "green" citizenship (Rancière 1995a; Žižek 1998) and,[10] therefore, authoritarianism. Pow and Neo (2015, 134) showed that sustainability discourses tend to follow certain commonalities (as established by international networks of policy experts, professional planners, architects or engineers), and are strongly driven by entrepreneurial objectives and ideals of ecological modernization flooding a city's every corner (shopping mall publicity, metro and bus stations, street banners). These are ideas that, as I showed in Chapters 4 and 5, city dwellers consciously or unconsciously absorb during a normal daily commute, or by taking part in ordinary activities run by SGOs in their communities. Indeed, urban planning serves as a sort of practice for reading and rewriting the ideal imaginaries of a modern city. In China, policymakers make particular use of master plans to reinforce control and create a favorable outcome for themselves. As de Certeau (1984, 94) suggested, discourses of urbanism seek to "repress all the physical, mental and political pollutions that would compromise it". In the following lines, I make use of de Certeau's concepts of "strategy" and "tactics" to explore how Shanghai 2035 functions as a strategy by the state to keep power, by looking at how space is envisioned and regulated (Foucault's panopticon).

Shanghai 2035: building a "modern eco-socialist country"

According to de Certeau, all societies operate under a dual structure: on the one hand, the dominant system (e.g., China's authoritarian regime); and on the other, the users who are thought to be passive within this system (e.g., SGOs and citizens) (Demirpolat 2021). To maintain their overarching framework, those in power develop "strategies" which are then used and appropriated by individuals or groups. Yet, contrary to Foucault or Lefebvre's pessimistic approach to individuals' agency,[11] de Certeau (1984) argues that within the spaces of power, those who are characterized as weak, or passive can engage in "tactics", understood as an "art of the weak". In the previous chapters, I have argued that the party and the state are attempting to foreclose spaces for "tactics" to emerge. In the urban sphere, particularly in the Chinese context where the state exerts substantial influence over political and economic resources, a city performs as a strategic tool for "effecting state power" (de Certeau 1984 cited in Chen and Lin 2021). Here, the "police" order identified in Rancière's writing is particularly useful in recognizing the role of space (the city) in governing (Dikeç 2002). It shows that the process of governing is the result of the ways various actors (human and non-human) structure the way we perceive ourselves, one another, and our surroundings. In the city, buildings, SGOs or street propaganda are symbolic of such forms of domination. Urban planning is also guided towards bringing order.

In China, Master Plans (*zongti guihua* 总体规划) are recognized for their role in guiding planning (F. Wu 2015; Morand 2019; Qiyu 2019; Wang 2019). They have been functioning as key guidelines for planning since 1990, when the planning Act came into effect. Since then, central government has required all development projects to provide guidance for future development (F. Wu 2015, 54). Yet, because master plans are quite general about future development, their implementation has been contingent on the "understanding" of government officials and the plans are often used to promote the country's economic development (*ibid*). Shanghai has a long history of city planning—since long before the planning Act. By the beginning of the 20th century, Shanghai was the first city in China that faced the pressure of modern life, with contradictory urban scenes, and immigrants from the world as a result of past foreign settlements (1843; see Chapter 5, p. 117) and, later, the influx of Chinese fleeing political disturbances in neighboring provinces (1853). This increase in population and the need to overcome the longstanding spatial fragmentation between the European quarter and the indigenous areas led policymakers to prepare and develop the Greater Shanghai plan as early as 1927 (F. Wu 2015, 8). Although implementation of the plan was interrupted as a consequence of the Japanese invasion, urban planning reemerged after Japan's surrender in 1945 and master plans have been playing a big role in the city's development ever since.[12]

Shanghai 2035 is the State Council-approved master plan, published by the city of Shanghai, for the years 2017–2035. The plan lays the groundwork for the development of a "global city" model. The goals are ambitious, mixing promises about air quality with narratives on finance, trade, and technological innovation. The master plan envisions "an admirable city of innovation, humanity and sustainability as well as a modern socialist international metropolis with world influence".[13] Overall, the Metropole is seen a display of Xi's goal of developing China into a great modern socialist country.[14]

Acting as China's showroom to the world, Shanghai is a relevant case study to broaden our understanding of "contemporary urbanisms(s) in their multiplicity" (Gibas and Boumová 2020, p. 20). Growing out of a small fishing village, to the "Oriental Paris" after the port opening in the mid-nineteenth century, Shanghai is today recognized as one of the most innovative and enterprising metropolises in the world (Huang 2020). These past ten years, the city became a frontrunner in showcasing the country's sustainability actions through the development of multiple pilot-projects such as, for instance, wetland restoration projects, or the implemention of a mandatory household waste sorting system in 2019 (see Chapter 4). All these efforts aim to help Shanghai "play a pioneering role" and "take the lead" by becoming an "environment-friendly, economically-developed, culturally-diversified, safe and liveable city" (Shanghai Planning and Land Resource Administration 2018 cited in Den Hartog 2021).

As in other cities in the world that exist within a range of political settings, the development and sustainability agendas of Shanghai are increasingly intertwined with the neoliberal agenda. "Green" planning and design ambitions are translated into a "successful city" within the global urban network competition, meaning that Shanghai caught on to the need to "green" itself to ensure the city remained attractive for capital investors (Adscheid and Schmitt 2021). Unlike in liberal democracies, however, efforts to instate sustainability goals in China are a process of strong top-down development, as local governments endeavor to disseminate and implement the goals set at a central political level. Consequently, more than any other city in the Mainland, Shanghai encapsulates Xi's paradoxical triple goal of expanding the role of the market, advancing state control, and establishing sustainable goals. As Isabella Weber (2020) put it, "China is found both to be neoliberal and to provide an alternative to neoliberal development". While this chapter does not aim to add to the extended literature on neoliberalism, it states that understanding the politics of "green" urbanization is essential to observe how different goals—political, economic, and social—affect everyday politics. In particular, it explores how sustainable urbanism must be recognized as a mismatch of local, national, and international interests, bringing competitive and authoritarian benefits—and not just aesthetic ones—to those in power (Krueger and Gibbs 2010).

The sustainable development narrative is at the center of Shanghai 2035 and perfectly encapsulates the reasons why China's way of governing needs to be analyzed not only as coercive and centralized, but as carefully choreographed through a variety of mechanisms including cooperative governance arrangements and "participatory" objectives related to sustainable growth. The imaginaries depicted in master plans such as Shanghai 2035 materialize the trend towards consensus built on the acceptance of the need for "green" development and public involvement. It is on this base that decision-making processes are being diffused through new actors, such as SGOs, and that alternative ideas and imaginations are excluded in favor of what Chinese leaders advance as "ecological civilization".

Public involvement in environmental governance

A closer look at the master plan and the consultation process shows that the project is guided by President Xi's instructions set out in the "14th Five-Year Plan" such as "opening doors and public participation in planning" (*kaimen zuo guihua* 开门做规划, in Chinese). As part of this trend, mechanisms have been included to ensure the participation of multiple parties—state, society, and private sector—from the establishment of the plan to its implementation. In July 2014, prior to the release of the plan, several of these mechanisms were used. A public survey was conducted, with nearly 14,000 questionnaires to better understand the public's values and goals in various fields.[15] The public's opinion was also collected through other methods such as forums, online platforms (Weibo, WeChat), or letters. Besides, under the requirements of the City Planning Law of the PRC (*Zhonghua renmin gongheguo chengxiang guihua fa* 中华人民共和国城乡规划法)—which includes article 54 requiring policy makers to publish plans for public consultation and supervision for a minimum period of 30 days—planning announcements were shared with the public.[16] The regulation does not, however, clarify either the rights and legal procedures related to the content and methods of participation, or its participants. It only stipulates that policymakers must publicly share their plans beforehand.

In this section, I draw on post-politics and my observations in the field to argue that such consultation/participation mechanisms and the growing involvement of SGOs water down public opposition and contestation (as discussed in Chapter 2). Here I am therefore defying more traditional scholarship that views citizen participation as the hallmark of democratization, and building on the work of scholars who assert that participation entails not only a democratic turn but may also be perceived as a threat to citizens' autonomy, as Blühdorn and Butzlaff (2020) suggest. Rather than empowering people, the sustainable imperative to tackle the environmental challenges or, new "green" urban panoramas, re-inscribes state power. Analyzing discourses on urbanization illustrates how the construction and performance of "green" targets place limitations on what we can say, and what kind of roles citizens have in the "ecological civilization" framework without ever bringing the status quo in question.

ZeroWaste and Garden leaders (see Appendix B), as well as several SGO workers with whom I spent time volunteering, shared their frustration regarding the implementation of these mechanisms for participation. They particularly emphasized the tremendous gap between the official rhetoric and their actual capacity for participation in decision-making processes. Indeed, the further restriction on their advocacy (as described in the previous chapters) considerably hindered possibilities for bottom-up approaches to emerge. Participation and public involvement are indeed a reality but are confined by the "green" consensus as exposed in this book. These growing expectations that SGOs have as they become more involved in governance processes are thus hampered by their inability to affect change. Indeed, their role has been strategically integrated in decision leaders' "soft" and "cooperative" governance strategies.

In-depth participation or disguised techno-managerialism?

It is clear that Chinese leaders have come to recognize the role of civil society in the last few years (Wu 2009; Johnson 2010; Teets 2014). Although levels of participation depend on local variables (L. Zhang et al. 2020), public participation has been stipulated in national policies. In his speech at the nineteenth Party Congress, Xi Jinping asserted that China must "expand the people's orderly political participation to see that in accordance with law they engage in democratic elections, consultations, decision-making, management, and oversight". As Ortmann (2016) argued, public pressures are necessary for monitoring and pushing for more stringent legislation at a local level. Taking Singapore as a case study, he shows how, despite significant restrictions on independent activism, a small environmental movement emerged as a response to a faulty environmental governance system, which was mainly centered on technological solutions and economic development. The growth of successful bottom-up campaigns led the state to accept greater involvement from the public if they functioned as a tool of the government.

Singapore's technocratic model of environmental governance has long functioned as a model for China (Curien 2017). The two countries strengthen their cooperation on environmental protection through several bilateral platforms such as, for instance, the Tianjin Eco-City which was designed to export Singapore's model to China, and which celebrated its tenth anniversary in 2017. The role that Singapore plays in China is not new. It is instructive here to recall Deng Xiaoping's admiration for the city-state's model during his Southern Tour in 1992 (Yang and Ortmann 2018). More recently, Masagos, Singapore's Minister for the Environment and Water Resources, highlighted in a Facebook post that: "Despite the different development paths of our countries, there are many areas for collaboration and mutual learning between Singapore and China. We reaffirmed that environmental cooperation is important for bilateral ties".[17]

China, like Singapore, has recognized the need to open spaces for participation, as Shanghai 2035 showcases. On the one hand, this responds to decentralization issues, while, on the other, the two countries share sustainability hurdles among the society at large. As Lo (2015, 152) pointed out: "Pure authoritarian environmentalism obviously does not exist in the ideal form in any context, just as pure neoliberal or democratic environmentalism does not exist". As such, there is a need to update the debate on China's environmental politics because governance mechanisms are increasingly scattered between different strategies which tend to escape our foundational ideas about what authoritarianism is. And, again, as my analysis is focused on Shanghai, I do not aim to defend the idea that we can generalize the findings outlined in this book on a national level. Yet, I believe such mechanisms will increasingly dictate the conditions of life in Chinese cities.

My empirical analysis shows that the CCP's rhetoric of an "ecological civilization", firm commitment to climate strategies, and efforts towards green urbanism are not solely environmental goals, but they represent a way to stabilize state priorities through a consensus-oriented and post-political approach. In this given environment, the potential for mobilizing SGOs is not only being curtailed by restrictive measures (e.g., registration hurdles, small funding) but by the fact that their ecological commitment is directed to complement the broader goals of the state. This limits their ability to go against the state's predominant narrative on sustainability, and leads to the depoliticization of SGOs because the "green" consensus strategically circumscribes the space allowed for action and debate in order to not disrupt the hegemonic discourse. Added to this, their "participation" in governance leads to a dispersion of responsibility for governing, as the supposed cooperative nature of state–society practices diverts attention from the state's role (and responsibility) in governance. Community spaces in the city play a significant role in such processes, as I showed with the case of ZeroWaste and, to a certain extent, Farming.

Even though urban planning is publicized as being more democratic by policymakers—because it is shaped by the assemblage of heterogeneous actors—I witnessed a far more complex reality on the ground. Because they create consensus while limiting dissent, the urban imaginaries advanced by SGOs promote the state's conception of sustainability and, thus, obfuscate the fact that the construction of such discourses entails acts of power and "policing". The "green" economy is increasingly used as a political mantra to advance the country's pathway towards becoming a "modern socialist country that is prosperous, strong, democratic, culturally advanced and harmonious" (Hu et al. 2021). Engaging with the concept of post-politics does indeed provide a critical perspective to assess how the imperative of sustainability is leading to a pluralization in governance, yet to the exclusion of any voices contesting the hegemonic consensus advanced by Chinese leaders. Within this rhetoric, any alternative voices that speak in favor of other possibilities, as encapsulated by the rise of environmental movements in the 2000s (see Chapter 2), are denounced and dismissed as irrational or irresponsible.

My claim here is that political leaders build *environmental authoritarianism* with the involvement of a multitude of actors, as explored previously, because the current environment has enabled them to establish a broad and vague consensus about what is acceptable and desirable (Neo 2021, 115). This is where the "eco-city (*shengtai cheng* 生态城)" and its attended imaginaries and aspirations play a major role. Taking stock of Shanghai 2035 and juxtaposing it with the reflections from my empirical work in Shanghai, in the following sections, I turn my attention to China's deepening embrace of sustainable discourses in a context of climate change. The main idea here is to enquire more deeply into the effects that such urban projects can have on our understanding of environmental governance and state–society relations in China.

In doing so, I will further uphold the argument that *environmental authoritarianism* means that the CCP pursues environmentalism for its own advantage. The appeal of environmental innovations, however, is not unique to Shanghai. Shenzhen's "Global Pioneer City", Hangzhou's "World Famous City", Wuhan's 2049 Master Plan, and Beijing's 2035 Master Plan similarly point to cities on a green quest to be recognized as hubs for environmental innovations. There is an urgent need to engage with a critical gaze regarding how other cities are being narrated to contribute to our understanding of post-politicization and environmental politics in China's different contexts.

Naturalizing technocratic logics in planning policy

In 2013, as of the Third Plenary Session of the 18th Communist Party of China Central Committee, the central government announced its intention to control the size of the population of its largest city. The government argued that it would redirect the rural population towards small to medium-sized cities, speeding up their development. Environmental concerns were not on the agenda at this point, but later, the narrative changed. As the master plan put it, population growth should be controlled because it is crystallizing environmental degradation. To respond to the "big city disease",[18] the State

Council announced it would limit Shanghai's population to 25 million by 2035.

Shanghai 2035 reframes population control as a solution to mitigate the contradiction between rapid population growth and resource and environment restrictions. Urban planners employ an environmental narrative to depoliticize access to the city by linking it to sustainable narratives. The limitation of access to the city for environmental purposes is a simple and striking example, among others, showing how sustainability is being mobilized to legitimize a variety of command-and-control planning practices, such as smart urbanism, space coordination, management of the population (e.g., size ceiling); or, as Kaika (2017) put it, to "vaccinate" the addressees of these policies against their underlying social, economic, and political consequences. Likewise, by taking care to optimize ecological land use and promote integrated utilization of space, the master plan also seeks to strengthen street space management and control (Shanghai 2035, p. 63). When reading the theories of de Certeau, it becomes clear that policies such as this one harbor an inherent politics of space, and that space is intrinsically and unequivocally political. According to de Certeau's theory on the production of space, stories rather than maps or numbers build people's connection to their communities. Imaginaries, then, function as a valuable addition to the planner's toolbox, especially in creating mechanisms to increase community engagement and develop a consensual and clear-cut view regarding what needs to be done and how (Hendriks 2009). Decision makers undertook unpopular policies such as eradicating illegal residences or closing illegal migrant schools as part of these "urban renewal" projects.[19]

Scholars like Fulong Wu or Federico Caprotti have widely documented the downsides of massive demolition of large parts of the city. As the latter put it, the climate urgency is pushing the "engineering of new urban environments, often along ecologically modernizing and technocratic lines" (Caprotti 2014b). As such, the emergence of eco-cities is constructed on an assemblage of discourses around the necessity of technological and infrastructural fix so that it becomes more resilient to the environmental challenges. Following the work of these scholars, Lei Ping (2019) questions the ongoing destruction in the name of "sustainability" which, according to her, is destroying the "lively, neighbourly, collective, street and community-based self-sufficient urban way of life exemplified by the historic Shanghai *longtang*".[20] Moving around the city one can witness how this destruction is increasingly being framed within the "green" consensus as construction, while demolition sites are "packaged" by big green posters.

I remember a particular episode strolling around an old community north of Minhang District as I returned from a ZeroWaste voluntary activity. Small children were playing in an ancient winding street, a woman was piling furniture inside a handcart. I was walking with a colleague with whom I had spent the morning preparing fridge magnets to publicize Shanghai's new waste disposal system, that would later be distributed in primary schools.

I asked my colleague: "Are these buildings being demolished?" "Yes", he responded. "Where are these people going, then?", I asked. "They will be relocated or compensated. It is a good thing. These buildings conditions are very poor", he told me confidently. Seeing I was a foreigner, one kid yelled "hello". "How are you?" (in English). I replied, and asked whether and when he was going to move. "Very soon", the kid answered. "Where will you go?", I asked. "We are moving to my aunt's place until we get relocated to a new apartment. I am sad because it's very far away from here".

This is a salient example of how "green" narratives hide the strong social and inequal nature of such consensus-oriented approach to sustainable development. Indeed, this sustainable imaginary is not for everyone. With a need to cap its population, foreign and low-skilled workers were also being pursued and expelled from the city. Since April 2017, Shanghai set up a classification for work permits with three categories (A, B, C). Based on their education, qualification, skills, age, expertise or proficiency in Mandarin, each international migrant working in China is given a score, and, as Shanghai 2035 states, priority is to be given to "high-caliber talent (*gaoduan ren* 高端人)" (Shanghai 2035, p. 53). At the start of 2017, some of my colleagues from Russia, Colombia, and Peru, teaching English in private schools, were slowly seeing their working visas being canceled. Some chose to enroll at a Chinese University to be able to stay, meaning they were working illegally, but could stay in China. Yet it became increasingly challenging for them to find a job. All these examples show how environmental goals can be used to "disguise" authoritarian modes of governing and/or particular political interests. The policy regarding population control and work permits was announced after several episodes of migrant evictions were reported in both Shanghai and Beijing.[21]

Figure 6.1. Environmental Propaganda in a bus station (left) and covering a construction site (right).

State hegemony, community, and the sustainable city

Another interesting feature of the master plan is the way that issues such as tackling pollution problems (e.g., reducing PM 2.5 emissions), increasing green spaces and forest coverage (e.g., for 90 percent of parks to be at a five-minute walking distance), developing more public surveys, increasing the weight of the cultural sectors (e.g., for the proportion of people working in the cultural sector to reach 10 percent of total employment), and so forth, are interchangeably displayed among less "popular" targets, such as more "bottom-line control". In the master plan, urban planners also expect citizens to be "law-abiding, credible and well-mannered" (Shanghai 2035, p. 4) while the city is to "Strictly follow central government's requirement" (Shanghai 2035, p. 28).

The urban population is more heterogeneous, mobile, and economically independent in urban settings, adding complexity to the "build community" campaigns (Bray 2006) launched by the CCP nationwide. Yet, the party-state seems determined to rectify the appearance of "selfish" minded individuals (Steele and Lynch 2013) or individualist sentiments that oppose its urbanization project. The CCP's efforts to compensate for a lack of *suzhi* 素质 (inner quality) attest to this claim (see Hsu 2017). As Laliberté and Lanteigne (2007) put it, the CCP is pushing towards the development of "self-governed" and "self-contained" individuals:

> In terms of identity, urban residents are expected to shift from identifying themselves with *danwei* to identifying themselves with community/*shequ*, a "*shequ* person". The identity change also involves attitudinal and behavioral changes … replace dependence (*yilai*) and submission to authority (in the sense of simply doing as one is told) with independence and active participation in community/*shequ*.

In their book, *The Politics of Community Building in Urban China*, Heberer and Göbel (2011, 33) argue that an increasing process of individualization hinders the state's attempts at community-building. As they explain (*ibid*): "Residential areas, which are seldom composed of people from the same *danwei* but of people with very different social attributes, now no longer constitute communities of solidarity, but rather conglomerations of strangers". According to the authors, the retreat of the state and a lack of social fabric capacity (characteristic of *danwei* units) increase instability in a time of explosive unemployment and social insecurity.

The retired volunteers with whom I spent time when doing observation fieldwork in communities in Shanghai were quite aware of this issue as they complained to me about the fact that "migrants" and young generations don't engage in the community activities (I alluded to this point in Chapter 4). As such, "the other groups" were cataloged as "bad" citizens for not participating in activities ranging from waste initiatives, greening the community gardens, or casual karaoke sessions. SGOs were particularly active in promoting views of "good" citizenship based on norms of the self-responsible, active citizen. Be it ZeroWaste or other SGOs such as BlueOcean, Garden, or Mouse (refer to Appendix B and Appendix C), all of the organizations I encountered during fieldwork advanced individuals as responsible actors in protecting their environment. Through the construction of self-responsibilization as the unifying signifier, SGOs explicitly brought individuals together within a broad alliance connecting environmentalism with the question of "good" citizenship. As such, they reinforced the idea that everybody, as individuals, must work towards the environmental protection slogan advanced by the municipal government in the city space or "eco-city" planning discourses (see Figure 6.2).

This means that the idea of citizenship advanced in the "eco-city" imaginary is a social product, a representation shared and circulated by numerous actors which is aimed at cultivating active and responsible citizens while rehabilitating a bounded community framework, whereas the party-state keeps its central role in community-building. Besides, the concept of citizenship gives new roles to individuals within a complex combination of mobilizing, politicizing, and depoliticizing techniques by "bordering" mobility, as gated communities increasingly become the norm (see Chapter 3). In such delimited spaces, "green" references are particularly used to advance what should be done, and who they (citizens) should be: "Instead of maintaining discipline by means of coercive state organizations like the police or public security agencies, policy-makers seek to stimulate the forces of self-discipline in tightly knit 'communities'" (Heberer and Göbel 2011, 13). This has been a crucial goal for the CCP:

Figure 6.2. Recycling and environmental carnival poster.

> We will strengthen the system for community governance, shift the focus of social governance to the community level, leverage the role of social organizations, and see that government's governance efforts on the one hand and society's self-regulation and residents' self-governance on the other reinforce each other.
>
> (Nineteenth Party Congress, p. 44)

By following SGOs in their daily practices, I could observe how this wider strategy is increasingly informed by the power of individuals and their roles in creating a "beautiful" and "green" China. Like during Mao's period, *verdurization* narratives are used to emphasize the importance of cooperation between citizens and local governments in solving problems with the ultimate objective of sustainable development. Shanghai's new recycling scheme is an excellent illustration of this strategy. Local leaders have imposed compulsory sorting on citizens since 2019 through state bureaucratization in neighborhood communities (*shequ*). Individuals are monitored daily by a multitude of actors—community leaders, volunteers, non-governmental organizations, or waste pickers—to ensure regulations are followed. Such cooperative mechanisms in communities create new spaces (similar to the *shequ*) where the state's implicit and explicit state-building efforts are implemented.

Post-political configurations of urban environments

An increasing number of scholars frame post-politics as an obstacle to tackling climate change in an effective and democratic way. But what would be the value of reflecting on such issues in China, given the authoritarian nature of its political regime? My observations in Shanghai point to the fact that state–society relations resemble what post-political literature describes as governance based on consensus and expert-oriented politics. Recognizing this allows us to grasp new trends in authoritarian urban spaces and provides us with the conceptual tools to investigate how they are being defined, ordered, transformed, and understood as common entities, such as the "sustainable city" or "ecological civilization". These concepts are, in my view, essential in order to decipher spaces of depoliticization. As I show in this book, *environmental authoritarianism* works in/into urban flows and urban dynamics. Examining the post-political configurations of Shanghai 2035 and associated urban imaginaries, for instance, provides a more nuanced point of analysis from which the heterogeneous relations (re)producing post-political narratives can be traced. Following on from the previous chapters where I showed how SGOs' involvement in community governance increasingly contributes to black-boxing environmental authoritarian processes, in this chapter I show how city planning enables us to further reveal, unmask, and make visible the varying rationalities at play behind the spread of "sustainable" narratives, and how these contribute to the creation and spread of post-political environments (Adscheid and Schmitt 2021). Thus, one should not ignore the power of the narrative of the city which asserts certain types of knowledge and power structures.

A closer look at the settings and arrangements for participation reveals the underlying post-political condition of sustainable practices in Shanghai, similar to the context outlined in Yunci Cai's and Yujie Zhu's work on heritage (Lam-Knott et al. 2019). Cai shows how heritage is instrumentalized, and used to diffuse local disputes against urban gentrification and touristification (Cai 2019, 107). As a complement to these primary studies, this book asserts that post-politics has become a resource mobilized by Chinese leaders to serve their agendas by implementing "disguised" but powerful tools for urban governance, while obfuscating spaces for disagreement. Although there is a good deal of progress towards increasing public participation, the process remains passive (Yue et al. 2019) as it stays deeply coordinated by the state. There are few opportunities for direct dialogues between the government and local communities. Moreover, command-and-control policies are orchestrated by SGOs, community leaders, and volunteers, who are encouraged to promote public participation mechanisms but are rarely, themselves, involved in decision-making processes.

As underlined in Shanghai 2035, consultation, transparency, and participation are essential components for implementing "ecological civilization". Yet, the relationship between top-down and bottom-up governance approaches is still unbalanced. As shown in my research, sustainable goals have indeed enabled new cooperative relationships to emerge, but the actions of SGOs and cooperative governance mechanisms are less about making an effort to effect change, than about the reconfiguring of social space and the transformation of the power relations that eliminate any possibility of going against the status quo. As with the well-documented development of "ecological civilization", Shanghai 2035 is another example of an "empty signifier" (Laclau 2005, 2017; Offe 2009) that provides the opportunity for different actors to engage in planning, but where, crucially, power relations are rendered opaque because appeals to consensus-based politics around a desirable goal make it difficult to argue against the project. According to Laclau (2017), it is the tension between state-imposed limits and the fact that it is impossible for excluded parties to be properly represented that leads to policy that is "empty" in practice. This is the power of master plans: they can mean whatever their interpreters want them to mean. Yet, as Gunder and Hillier (2016) put it, sustainability does offer an overarching narrative around which practice can be oriented or—as I have tried to show—depoliticized.

These increasing depoliticization techniques helps authorities to impose top-down decisions on citizens more discreetly. Still, such centricity of top-down power should not be understood as simplistic; we must not neglect the complexity of the various interests at play. Indeed, various SGOs are competing for recognition or advancing different agendas, and local and national political actors are also competing to advance their own interests. Equally, market and entrepreneurial-oriented discourses also advance a multitude of ways of working and imagining the future of the city. Nevertheless, such competition is designed around the promotion of certain views of sustainability, which elevate certain spaces and actors over others.

Everyday resistances

In this final section, I will briefly explore in more detail my statement about the irreversibility of the post-political condition by focusing on de Certeau's concept of "tactics". As mentioned above, de Certeau distinguishes between the "strategies" of the powerful and the "tactics" of the "weak" to resist those strategies. The French scholar thus presents urbanism as the ultimate strategy used by those in power seeking to dominate and control space and people's everyday activities. Throughout this book, I have tried to showcase how environmentalism has become a strategy for the CCP to assert its power in urban areas such as Shanghai. A strategy to "conduct the conduct" of Chinese citizens (Foucault). Yet, although I came to realize how strong this "green" consensus is, I still think it is crucial to not underestimate how the complexity of today's global cities conceals potential for "tactics" to emerge despite the regime's attempts to reinforce discipline through various practices, whether authoritarian or more cooperative.

This is where I find de Certeau's desire to conceptualize resistance, while not neglecting how pervasive the power of the state is, very promising. It adds substance to Foucault's understanding of power relations and technologies of power, which primarily focuses on imposing power on the individual. de Certeau shows that "everyday" activities, sometimes qualified as "misconduct" or "negligence", are in fact "tactical" responses to specific strategies of urban planning. Wandering around the city, I witnessed dozens of instances of such "tactics", be it people jeopardizing the recycling facilities, refusing to put their bags in the metro scanner machines, driving cars without license plates, or tagging propaganda posters and walls, "pedestrian" level (Beaumont 2020) tactics resist the disciplinary disposition of space. One needs only to open one's eyes to the Metropole's rich history of dissent, resistance, and everyday action to witness how the city is not just a site for those in power to assert their legitimate disciplinary authority.

Another interesting example of this is the way in which street food vendors kept returning after being chased away by the police in some city center streets. Night after night, I witnessed this game of cat-and-mouse, and night after night, the street vendors kept coming back again and again. In view of the increasing reassertion of power by the regime, I believe it is at the street level that scholars should be looking for resistance in the years to come.

Conclusion

The continuous push for economic development and the inherent damage this does to the environment faced growing disapprobation from Chinese civil society and the international community, and these were interpreted by many as the perfect recipe for a social upsurge (Chang 2010). Moreover, many have argued that an authoritarian regime is inherently unfit to respond to such major environmental issues, and this will increase the challenges faced by the CCP in future. Still, as explored empirically throughout the book, there are increasing signs that the party-state is managing to respond to challenges to its environmental legitimation in urban areas such as Shanghai. Moreover, primary evidence demonstrates that recent development is helping the CCP to regain legitimacy both at national and international levels. Events appear to be moving in the opposite direction to what many analysts expected.

If we take the concept of *environmental authoritarianism* even further, this chapter argues that environmental protection and sustainability are used by the CCP to further anchor its grip on power. Analyzing the case of urban planning, this chapter helps interpret recent developments in environmental politics and governance and, in particular, sheds light on the adaptations made by authoritarian regimes in responding to the environmental threat through city planning and attendant imaginaries. Broadly speaking, the chapter claims that China is continuing to follow a path of accelerated development, and at the wheel of a growing hyper-functionalist urban machine (Curien 2014), and this reinforces the need to assess the new methods of operation of authoritarian regimes. Taking Shanghai 2035 as an example, this chapter concludes that China's new governance model is inscribed in what some scholars have framed a "post-political" condition. As seen above, the master plan reduces politics to management and administration, while experts dominate the decision-making, and the state defuses conflict through cooperation with SGOs and other "responsible" partners. As such, SGOs initiatives are co-opted into a framework based on cooperation and dialogue, while the environment is used as a tool for consensus-building to develop pragmatism and eliminate ideological debates and antagonism. The Metropole's aim is to achieve economic growth, a "global city" status, develop security in the face of threats, and improve environmental sustainability, but in this urban machinery, the market economy and globalization are seen as irreversible facts.

Therefore, while this chapter demonstrates the state's ability to manipulate the current environmental situation in order to pursue specific goals and justify a range of policies that, as this chapter has tried to show, have different and contradictory objectives. Future research should not neglect to explore the inconsistencies and constant challenges of environmental policy in China. Urban planning processes that focus largely on satisfying a small portion of Shanghai's well-off population could create frictions among different strands of the population in the near future. Yet, as shown through Shanghai 2035 case study, the ambitious goal of limiting access to the city—some Chinese scholars already warned that Shanghai's population has already passed, according to scholar Yan Song, director of the University of North Carolina's programme on Chinese cities, the pointless target of 25 million[22]—is less important than the political effect that such a policy creates among the Chinese population and the international scene. As Zinda (2018, 67) states:

> In visions of ecological civilisation, from now on China's natures will be defined, built, and maintained by humans, in the vision of the CCP. This is no longer a matter of conquering nature, but of establishing flexible mechanisms of monitoring and response around complex and unpredictable social and environmental processes.

To cut a long story short, given that the current book reflects the current role of environmental organizations in China's urban governance processes, it was important to include a chapter that zooms further in on the broad picture of the party-state's dominant strategies regarding the environmental cause. On the one hand, this chapter highlights how SGOs position themselves in the party-state's main vision and, on the other, it depicts how state–society relationships could evolve in the near future. Besides, this chapter further highlights that the Chinese leadership is not immune to certain influences such as the consequence of social and political changes in late modernity (e.g., individualization, technological progress, neoliberal ideology). Therefore, it reinforces the role that the new forms of hybrid organizations identified in Chapter 5 could take in the future.

Nevertheless, we should not forget that the focus of this book is on Shanghai and these issues would certainly be approached differently in other Chinese regions. Therefore, I do not aim to argue that the approach to *environmental authoritarianism* related here can be generalized, or that it reflects a wider Chinese approach to environmental governance. Still, given the importance of Shanghai and its international dimension, this is an important point of departure to further assess how the CCP's adaptation strategies coexist within neoliberal forms and, particularly, the way the CCP capitalizes on climate change knowledge, techniques, and logic to address local challenges and persuade citizens to commit to the "ecological civilization" path.

Notes

1 For instance, a recent article in the *South China Morning Post* indicates that China has been issuing preferential policies for taxation and foreign investment to narrow its gap with Hong Kong. This move could affect Hong Kong's leading role in the gold market in favour of Shenzhen. These policies are used strategically to weaken Hong Kong's position in China's Greater Bay Area plan. See Karen Yeung (20 August 2019) Are Hong Kong's protests crushing the city's role in China's Greater Bay Area plan?. *South China Morning Post*. https://www.scmp.com/economy/china-economy/article/3023615/are-hong-kongs-protests-crushing-citys-role-chinas-greater (accessed 23 August 2019).

2 For more information, see http://www.urbanchinainitiative.org/en/research/usi.html.

3 Refer to (20 December 2016) 加强高校意识形态管控，习近平到底怕什么？ [Strengthening ideological control over universities, what does Xi Jinping actually fear?]. *Voice of America in Chinese*. https://www.voachinese.com/a/io-20161219-china-higher-education-brainwash/3641994.html

4 Many scholars have been observing how, for instance, China has been using a narrative of solidarity against Western colonialism in African countries (Callahan 2012).
5 The term is used by the author to describe the Communist regime promotion of "garden landscapes" which followed the Soviet Union urban planning model in the early 20th century. *Verdurisation* campaigns (tree planting, park building, renovating pomegranate gardens, etc.) were promoted to enhance the state's legitimacy and extend state control over public spaces.
6 See the website of the National People's Congress (NPC) of the People's Republic of China, available at http://www.npc.gov.cn/englishnpc/xjptgoc/xjptgoc.shtml (accessed 27 April 2022).
7 See Chinese Propaganda Posters gallery, environment collection, available at https://chineseposters.net/themes/environment (accessed 12 January 2022).
8 Posters from the 1970–80s that focused on the environment can be found here: https://chineseposters.net/themes/environment.php (accessed 3 December 2018).
9 Ben Turner (22 January 2020) Chinese city apologises for "naming and shaming" residents wearing pyjamas in the streets. *Telegraph.* https://www.telegraph.co.uk/news/2020/01/22/chinese-city-apologises-naming-shaming-residents-wearing-pyjamas/ (accessed 25 September 2019).
10 Rancière uses the concept of "police" not to refer to folks in uniform or the repressive arm of the state, but to the order of bodies that define the allocation of "ways of doing, ways of being, and ways of saying", in a particular society to maintain social order. Rancière's "police" refers to the hierarchical social formations (governing and administrative machinery) that individuals encounter in their everyday lives.
11 French sociologist Lefebvre developed a pessimistic approach to understanding modern consumer society and the impact of the universalization of exchange-value on society and culture, leaving little space for human agency. See Demirpolat 2021.
12 This information comes from Fulong Wu's book *Planning for Growth* (2015), highly recommended reading for any readers interested in the history of urban and regional planning in China.
13 The Shanghai Master Plan 2035 is available at http://www.shanghai.gov.cn/newshanghai/xxgkfj/2035004.pdf (see p. 20).
14 Based on the author's own observation during a field visit in late March 2017, Shanghai is displayed as the ultimate role model for what to achieve in remote areas such as Urumqi.
15 See Shanghai City Master Plan Report (2017–2035).
16 See the Law of the People's Republic of China on Urban and Rural Planning.
17 Masagos Zulkifli Facebook post, 20 September 2018, retrieved from Masagos's personal Facebook page (accessed 26 November 2018).
18 Characterised as an overcrowded, polluted city with too many people living in it. See Xinhua News Agency (1 May 2018) "*Shànghǎi gōngbù wèilái 18 nián chéngshì zǒngtǐ guīhuà jiāng nǔlì jiànchéng zhuóyuè de quánqiú chéngshì* (Shanghai announces the city's master plan for the next 18 years. It will strive to build a remarkable global city)", *XinhuaNet*, available at http://sh.xinhuanet.com/2018-01/05/c_136872844.htm (accessed 14 November 2018).
19 The claims I am advancing here are based on a conference I attended in Shanghai, in March 2017. During this event, the leader of a SGO providing English classes to migrant schools argued that informal schools, which were previously tolerated, were being closed as a result of profit-oriented real estate development plans.
20 *Longtang*, sometimes called *lilong*, is a neighbourhood of lanes populated by small clusters of houses that were built in the early 20th century.
21 Zhong Changqian (3 April 2017) Shanghai Migrants Face Eviction From Their Houseboats: *Sixth Tone.* http://www.sixthtone.com/news/2148/shanghai-migrants-face-eviction-from-their-houseboats/ (accessed 21 April 2018); Helen Roxburgh

(19 March 2018) China's radical plan to limit the populations of Beijing and Shanghai. *The Guardian.* https://www.theguardian.com/cities/2018/mar/19/plan-big-city-disease-populations-fall-beijing-shanghai (accessed 26 June 2018).

22 On this subject see Rémi Curien (10 January 2016) "Mondes urbains Chinois/services essentiels en réseaux et fabrique urbaine en Chine: verdire le développement accéléré?". *Urbanités,* available at http://www.revue-urbanites.fr/services-essentiels-en-reseaux-et-fabrique-urbaine-en-chine-verdir-le-developpement-accelere/ (accessed 10 January 2019).

References

Adscheid, Toni, and Peter Schmitt. 2021. "Mobilising Post-Political Environments: Tracing the Selective Geographies of Swedish Sustainable Urban Development". *Urban Research & Practice* 14 (2): 117–137.

Ahlers, Anna L., and Yongdong Shen. 2018. "Breathe Easy? Local Nuances of Authoritarian Environmentalism in China's Battle against Air Pollution". *The China Quarterly* 234: 299–319.

Beaumont, Matthew. 2020. *The Walker: On Finding and Losing Yourself in the Modern City.* Verso Books.

Beeson, Mark. 2016. "Environmental Authoritarianism and China". In *The Oxford Handbook of Environmental Political Theory,* edited by T. Gabrielson, *et al.*, 520–532. Oxford University Press.

Blühdorn, Ingolfur, and F. Butzlaff. 2020. "Democratization beyond the post-democratic turn: towards a research agenda on new conceptions of citizen participation". *Democratization* 27 (3): 369–388.

Bray, David. 2006. "Building "Community": New Strategies of Governance in Urban China". *Economy and Society* 35 (4): 530–549.

Brehm, Stefan and Jesper Svensson. 2017. "A fragmented Environmental State? Analysing Spatial Compliance Patterns for the Case of Transparency Legislation in China". *Asia-Pacific Journal of Regional Science* 1 (2): 471–493.

Cai, Yunci. 2019. "'Connecting Emotions through Wells': Heritage instrumentalisation, civic activism, and urban sustainability in Quanzhou, China". In *Post-Politics and Civil Society in Asian Cities,* 106–120. Routledge.

Callahan, William A. 2012. "Sino-Speak: Chinese Exceptionalism and the Politics of History". *Journal of Asian Studies* 71 (1): 33–55.

Caprotti, Federico. 2014a. "Critical Research on Eco-Cities? A Walk through the Sino-Singapore Tianjin Eco-City, China". *Cities* 36: 10–17.

Caprotti, Federico. 2014b. "Eco-urbanism and the Eco-city, or, Denying the Right to the City?". *Antipode* 46 (5): 1285–1303.

Chang, Gordon G. 2010. *The Coming Collapse of China*. Random House.

Chen, Siyu and Jian Lin. 2021. "Making with Shenzhen (Characteristics)—Strategy and Everyday Tactics in a City's Creative Turn". *Sustainability* 13 (9): 4923.

Cheshmehzangi, Ali. 2016. "China's New-Type Urbanisation Plan (NUP) and the Foreseeing Challenges for Decarbonization of Cities: A Review". *Energy Procedia* 104, 146–152.

Chu, Cecilia L. 2015. "Aspects of Urbanization in China: Shanghai, Hong Kong, Guangzhou". *Planning Perspectives* 30 (4): 665–667.

Chu, Yin-wah. 2020. "China's New Urbanization Plan: Progress and Structural Constraints". *Cities* 103, 102736.

Curien, Rémi. 2014. "Chinese Urban Planning. Environmentalising a Hyper-Functionalist Machine?" *China Perspectives* (3): 23–31.

Curien, Rémi. 2017. "Singapore, a Model for (Sustainable?) Urban Development in China". *China Perspectives* (1): 25–35.

de Certeau, Michel. 1984. "Walking in the City". In *The Practice of Everyday Life*, 91–110. University of California Press.

de Jong *et al.* 2016. "Eco City Development in China: Addressing the Policy of Implementation Challenge". *Journal of Cleaner Production* 134, 31–41.

Demirpolat, Aznavur. 2021. "Understanding De Certeau's Concepts of Strategy and Tactics in Relation to the Educational Policy Analysis". *Educational Policy Analysis and Strategic Research* 16 (3): 350–362.

Den Hartog, Harry. 2021. "Engineering an Ecological Civilization Along Shanghai's Main Waterfront and Coastline: Evaluating Ongoing Efforts to Construct an Urban Eco-Network". *Frontiers in Environmental Science.* https://www.frontiersin.org/articles/10.3389/fenvs.2021.639739/full.

Dikeç, Mustafa. 2002. "Police, Politics, and the Right to the City". *GeoJournal* 58 (2): 91–98.

Eberhardt, Christopher. 2015. "Discourse on Climate Change in China: A Public Sphere without the Public". *China Information* 29 (1): 33–59.

Fowler, Edmund P. 1992. *Building Cities that Work*. McGill-Queen's University Press.

Gare, Arran. 2017. "From 'Sustainable Development' to 'Ecological Civilization': Winning the War for Survival". *Cosmos and History* 13 (3): 130–153.

Gibas, Petr, and Irena Boumová. 2020. "The Urbanization of Nature in a (Post) Socialist Metropolis: An Urban Political Ecology of Allotment Gardening". *International Journal of Urban and Regional Research* 44 (1): 18–37.

Gibert, Marie, and Juliette Segard. 2015. "Urban Planning in Vietnam: A Vector for a Negotiated Authoritarianism?". *Justice Spatiale—Spatial Justice* 8. http://www.jssj.org/article/lamenagement-urbain-au.

Gunder, Michael, and Jean Hillier. 2016. *Planning in Ten Words or Less: A Lacanian Entanglement with Spatial Planning*. Routledge.

Hamnett, Chris. (2020) "Is Chinese Urbanisation Unique?" *Urban Studies* 57 (3): 690–700.

Heberer, Thomas, and Christian Göbel. 2011. *The Politics of Community Building in Urban China*. Routledge.

Hendriks, Carolyn M. 2009. "Policy Design without Democracy? Making Democratic Sense of Transition Management". *Policy Sciences* 42 (4): 341–368.

Hoffman, Lisa. 2011. "Urban Modeling and Contemporary Technologies of City-Building in China: The Production of Regimes of Green Urbanisms". In *Worlding Cities: Asian Experiments and the Art of being Global*, edited by A. Roy and A. Ong, 55–76. Blackwell Publishing Limited.

Hsu, Carolyn L. 2017. *Social Entrepreneurship and Citizenship in China: The Rise of NGOS in the PRC.* Routledge.

Hu, Angang, Yilong Yan, Xiao Tand, and Shenglong Liu. 2021. "Conclusion: The Mission of the Communist Party of China". In *2050 China*, 89–90. Springer.

Huang, Zhongjing. 2020. *Introduction to "the Shanghai Model"*. Taylor & Francis.

Jia, Kai, and Shaowei Chen. 2019. "Could Campaign-Style Enforcement Improve Environmental Performance? Evidence from China's Central Environmental Protection Inspection". *Journal of Environmental Management* 245, 282–290.

Johnson, Thomas. 2010. "Extending environmental governance: China's environmental state and civil society". University of Glasgow. PhD Dissertation.
Kaika, Maria. 2017. ""Don't Call Me Resilient Again!": The New Urban Agenda as Immunology … or … What Happens When Communities Refuse to be Vaccinated with "Smart Cities" and Indicators". *Environment and Urbanization* 29 (1): 89–102.
Kostka, Genia, and Jonas Nahm. 2017. "Central-Local Relations: Recentralization and Environmental Governance in China". *The China Quarterly*, 231, 567–582.
Kostka, Genia, and Chunman Zhang. 2018. "Tightening the Grip: Environmental Governance under Xi Jinping". *Environmental Politics* 27 (5): 769–781.
Krueger, Rob, and David Gibbs. 2010. "Competitive Global City Regions and 'Sustainable Development': An Interpretive Institutionalist Account in the South East of England". *Environment and Planning A* 42 (4): 821–837.
Laclau, Ernesto. 2005. "Populism: What's in a Name?" In *Populism and the Mirror of Democracy*. Verso.
Laclau, Ernesto. 2017. "Why Do Empty Signifiers Matter in Politics?" In *Deconstruction*, 405–413. Routledge.
Laliberté, André, and Marc Lanteign. 2007. *The Chinese Party-State in the 21st Century: Adaptation and the Reinvention of Legitimacy*. Routledge.
Lam-Knott, Sonia, Creighton Connolly, and Kong Chong Ho. 2019. *Post-Politics and Civil Society in Asian Cities: Spaces of Depoliticisation*. Routledge.
Lawreniuk, Sabina. 2021. "'A War of Houses and a War of Land': Gentrification, Post-Politics and Resistance in Authoritarian Cambodia". *Environment and Planning D: Society and Space* 39 (4): 645–664.
Lefebvre, Henri. (2000) *La Production de l'Espace*. Economica.
Lendvai-Bainton, Noemi, and Dorota Szelewa. 2021. "Governing New Authoritarianism: Populism, Nationalism and Radical Welfare Reforms in Hungary and Poland". *Social Policy & Administration* 55 (4): 559–572.
Li, Guoping and Hong Zhou. 2016. "The Systematic Politicization of China's Stock Markets". *Journal of Contemporary China* 25 (99): 422–437.
Li, Huifeng, and Martin de Jong. 2017. "Citizen Participation in China's Eco-City Development. Will 'New-Type Urbanization' Generate a Breakthrough in Realizing it?" *Journal of Cleaner Production* 162: 1085–1094.
Lo, Kevin. 2015. "How Authoritarian is the Environmental Governance of China?" *Environmental Science and Policy* 54, 152–159.
Logan, John R. 2018. "People and Plans in Urbanising China: Challenging the Top-Down Orthodoxy". *Urban Studies* 55 (7): 1375–1382.
Lu, Hanchao. 2017. "Shanghai Flora: The Politics of Urban Greening in Maoist China". *Urban History*, 1–22.
Ma, Yun. 2017. "Vertical Environmental Management: A Panacea to the Environmental Enforcement Gap in China?" *Chinese Journal of Environmental Law* 1 (1): 37–68.
McKee, Kim. 2009. "Post-Foucauldian Governmentality: What Does it Offer Critical Social Policy Analysis?". *Critical Social Policy* 29 (3): 465–486.
Miao, Bo, and Graeme Lang. 2015. "A Tale of Two Eco-Cities: Experimentation under Hierarchy in Shanghai and Tianjin". *Urban Policy and Research* 33 (2): 247–263.
Morand, Lucie. 2019. *Le plan: outil générateur de stratégies d'urbanisation durable. Cas d'étude sur Xiamen en Chine*. Université Paris-Est.

Mössner, Samuel. 2016. "Sustainable Urban Development as Consensual Practice: Post-Politics in Freiburg, Germany". *Regional Studies* 50 (6): 971–982.

Neo, Harvey. 2021. "The Post-Politics of Environmental Engagement in Singapore". In Environmental Movements and Politics of the Asian Anthropocene, edited by Paul Jobin, Ming-sho Ho, andMichael Hsin-huang Hsiao, 109–138. ISEAS—Yusof Ishak Institute.

Offe, Claus. (2009) "Governance: An 'Empty Signifier'?" *Constellations* 16 (4): 550.

Ortmann, Stephan. 2016. *Environmental Governance under Authoritarian Rule Singapore and China.* Southeast Asia Research Centre (SEARC) Working Paper Series 189.

Ping, Lei. 2019. "Demolition of a Distinctive Chinese Habitus: Controversies of Urban Sustainability in Shanghai". *Environment: Science and Policy for Sustainable Development* 61 (6): 4–17.

Pow, C.-P. 2012. "China Exceptionalism? Unbounding Narratives on Urban China". In *Urban Theory Beyond the West*, 62–79. Routledge.

Pow, C.-P. 2018. "Building a Harmonious Society through Greening: Ecological Civilization and Aesthetic Governmentality in China". *Annals of the American Association of Geographers* 108 (3): 864–883.

Pow, C. P., and Harvey Neo. 2013. "Seeing Red over Green: Contesting Urban Sustainabilities in China". *Urban Studies* 50 (11): 2256–2274.

Pow, C.-P., and Harvey Neo. 2015. "Modelling Green Urbanism in China". *Area 47* (2): 132–140.

Qiyu, Tu. 2019. "Shanghai Master Plan 2017–2035: 'Excellent Global City'", *Tous urbains* (3): 58–63.

Rancière, Jacques. 1995a. *La mésentente.* Éditions Falilée.

Rancière, Jacques. 1995b. *On the Shores of Politics.* Verso.

Ren, Xuefei. 2012. "'Green' as Spectacle in China". *Journal of International Affairs*, 65 (2): 19–30. https://www.academia.edu/50854721/Xuefei_Ren_Green_as_spectacle_in_china.

Rosol, Marit, Vincent Béal, and Samuel Mössner. 2017. "Greenest cities? The (Post-) Politics of New Urban Environmental Regimes". *Environment and Planning A* 49 (8): 1710–1718.

Shen, Wei, and Dong Jiang. 2021. "Making Authoritarian Environmentalism Accountable? Understanding China's New Reforms on Environmental Governance". *Journal of Environment & Development* 30 (1): 41–67.

Shi, Xiaojin, and Jinxi Wu. 2018. "Politicization of Global Warming and Energy Restructuring in China". *IOP Conference Series: Earth and Environmental Science* 151 (1).

Shiuh-Shen, Chien. 2013. "Chinese Eco-Cities: A Perspective of Land-Speculation-Oriented Local Entrepreneurialism". *China Information* 27 (2): 173–196.

Smith, R. 2020. *China's Engine of Environmental Collapse.* Pluto Press.

Steele, Lisa G., and Scott M. Lynch. 2013. "The Pursuit of Happiness in China: Individualism, Collectivism, and Subjective Well-Being during China's Economic and Social Transformation". *Social Indicators Research* 114 (2): 441–451.

Swyngedouw, Erik. 2009. "The Antinomies of the Postpolitical City: In Search of a Democratic Politics of Environmental Production". *International Journal of Urban and Regional Research* 33 (3): 601–620.

Swyngedouw, Erik. 2010. "Apocalypse Forever?: Post-Political Populism and the Spectre of Climate Change". *Theory, Culture and Society* 27 (2): 213–232.

Swyngedouw, Erik. 2011. "Whose Environment? The End of Nature, Climate Change and the Process of Post-Politicization". *Ambiente & Sociedade* 14 (2): 69–87.

Swyngedouw, Erik. 2019. *Promises of the Political: Insurgent Cities in a Post-Political Environment*. MIT Press.

Swyngedouw, Erik, and Japhy Wilson. 2014. "There is no Alternative". In *The Post-Political and Its Discontents*, 299–312. Edinburgh University Press.

Teets, Jessica C. 2014. *Civil Society under Authoritarianism: The China Model*. Cambridge University Press.

Termine, Constanza, and Daniele Brombal. 2018. "Ecological Civilisation for China's Urban Sustainability: A Review of Normative Criteria". *Monde chinois* (4): 59–70.

Thomsen, André, Frank Schultmann, and Niklaus Kohler. 2011. "Deconstruction, Demolition and Destruction". *Building Research and Information* 39 (4): 327–332.

Turhana, Ethemca, and Arif Cem Gündoğan. 2017. "The Post-Politics of the Green Economy in Turkey: Re-Claiming the Future?" *Journal of Political Ecology* 24 (1): 277–295.

Verdini, Giulio. 2015. "Is the Incipient Chinese Civil Society Playing a Role in Regenerating Historic Urban Areas? Evidence from Nanjing, Suzhou and Shanghai". *Habitat International* 50, 366–372.

Wang-Kaeding, Heidi. 2021. *China's Environmental Foreign Relations*. Routledge.

Wang, Lan. 2019. "Planning for Urban Visions: The Case of the Shanghai 2040 Master Plan". In *Handbook on Urban Development in China*, 22–35.

Weber, Isabella. (2020) "Origins of China's Contested Relation with Neoliberalism: Economics, the World Bank, and Milton Friedman at the Dawn of Reform". *Global Perspectives* 1 (1).

Wei, C. X. George. (2008) "Politicization and De-Politicization of History: The Evolution of International Studies of the Nanjing Massacre". *The Chinese Historical Review* 15 (2): 242–295.

Wells, Tamas, and Vanessa Lamb. 2021. "The Imaginary of a Modern City: Post-Politics and Myanmar's Urban Development". *Urban Studies*.

Wilson, Japhy, and Erik Swyngedouw. 2014. *The Post-Political and Its Discontents: Spaces of Depoliticisation, Spectres of Radical Politics*. Edinburgh University Press.

Wu Fulong 2009. *Environmental Activism and Civil Society Development in China*. Harvard-Yenching Institute Working Paper Series.

Wu Fulong. 2015. *Planning for Growth: Urban and Regional Planning in China*. Routledge.

Wu Fulong. 2016. "State Dominance in Urban Redevelopment: Beyond Gentrification in Urban China". *Urban Affairs Review* 52 (5): 631–658.

Wu Fulong. 2020. "Adding New Narratives to the Urban Imagination: An Introduction to 'New Directions of Urban Studies in China'". *Urban Studies* 57 (3): 459–472.

Wu, Fulong, and Fangzhu Zhang. 2008. "Planning the Chinese City:Governance and Development in the Midst of Transition". *The Town Planning Review* 79 (2): 149–156.

Wu, Shufang. 2015. "Politicisation and De-Politicisation of Confucianism in Contemporary China: A Review of Intellectuals". *Issues and Studies* 51 (3): 165–205.

Xu, Jilin. 2018. *Rethinking China's Rise: A Liberal Critique*. Cambridge University Press.

Yang, Kai, and Stephan Ortmann. 2018. "From Sweden to Singapore: The Relevance of Foreign Models oor China's Rise". *The China Quarterly* 236, 946–967.

Yue, Wenze, Qiu Shuangshuang, Zhang Qun, and Yang Huake. 2019. "新媒体时代城市规划的公众参与：话题演进与情感判别 [Public Participation in Urban Planning in the New Media Era: Topic Evolution and Sentiment Analysis]". *Journal of Zhejiang University* 49 (4): 105–118.

Zhang, Fangzhu, and Fulong Wu. 2021. "Performing the Ecological Fix under State Entrepreneurialism: A Case Study of Taihu New Town, China". *Urban Studies.*

Zhang, Lin *et al.* 2020. "Heterogeneity of Public Participation in Urban Redevelopment in Chinese Cities: Beijing versus Guangzhou". *Urban Studies* 57 (9): 1903–1919.

Zinda, John A. 2018. "Managing the Anthropocene The Labour of Environmental Regeneration". *Made in China* 3 (4): 62–67. http://www.chinoiresie.info/PDF/Made-in-China-04-2018.pdf.

Žižek, Slavoj. 1998. "For a Leftist Appropriation of the European Legacy". *Journal of Political Ideologies* 3 (1): 63–78.

7 Concluding thoughts

Environmental authoritarianism: from theory to practice

Introduction

The different chapters of this book have tried to reflect on a set of presumptions (mostly from the Western world), that awareness of environmental issues and the increasing number of SGOs (Social Good Organizations, see Chapter 1, p. 000) should be conducive to democracy or, at least lead to a weakening of China's one-party rule. By empirically assessing the way that state–SGO relations grew in Shanghai from 2016 to 2018, this book challenges these presumptions. Throughout the various chapters, I show that the attitude of China's leaders towards the environment has developed from a topic to be avoided to something that is desired. For although it was long seen as the world's worst environmental offender, China moved in the opposite direction by taking a "green" turn. The environmental question is not to be seen as a "burden" for the CCP's image, but as an important asset. The CCP's novel approach in articulating the environment is well reflected in the concept of "ecological civilization", which, as shown in Chapter 2, was elevated to unprecedented heights.

To reflect on China's "green" turn, I used *environmental authoritarianism* as a theoretical framework to challenge the dogma on authoritarianism as a model which is driven by top-down, non-participatory approaches, further reinforced by Xi Jinping's centrally controlled state machinery. Developing my argument through my empirical data, I stress that the CCP's approach to environmental governance encompasses a dual use of coercive and more "participatory" mechanisms. One of the major objectives of this book has been to provide a more nuanced understanding of the use of environmentalism as a rhetoric that reinforces authoritarianism. It has centered on three ideas that all challenge the standard understanding of environmental governance approaches in authoritarian regimes.

The first idea is that there is currently no room for a contentious environmental movement to develop in Chinese urban spaces. This idea results from the current consensus on the need to protect the environment, which resists the antagonistic dimension (Mouffe 2016) of environmental politics and creates a hegemonic order aimed at fulfilling the state's vision of "sustainable"

DOI: 10.4324/9781003231325-7

development. The overall recognition of the need to tackle environmental issues in urban or environmental governance marks the emergence of a "green" consensus whereby politics is replaced by hard and soft technologies of administration aimed at reshaping state–society relations.

Second, this "green" consensus opens new opportunities for the state to shape alternative forms of governance which incorporate new social actors, such as SGOs, in the arena of governing (Swyngedouw 2019, 13). As I show in Chapter 3, Shanghai is a pioneer in using SGOs in the provision of social services, and it therefore functions as a good ground to deconstruct the dynamic ways by which the state is developing a more cooperative approach to governance. It shows that the social governance model proposed in the report to the nineteenth National Congress of the Communist Party of China—with the aim of creating a social governance pattern of joint construction, co-governance, and sharing (*dazao gong jian gong zhi gongxiang de shehui zhili geju* 打造共建共治共享的社会治理格局)[1]—is being followed and implemented at the local level and non-state actors are increasingly used as governance tools to reinforce the state's legitimacy, recentralize decision-making, and eliminate dissensus. This is leading to a shift from fragmentation to a consolidation, leaving less room for contentious participation (Fu and Distelhorst 2018), one of the main reasons being that SGOs have come to be increasingly integrated into a legal regulatory system specifically designed to offer opportunities to those seen as suitable pipelines for development, and foreclose opportunities to anyone else. Restricted in their action and increasingly dependent on the state for funding, SGOs need to keep "in line" to survive.

Third, the "consensual scripting"[2] of environmentalism, as a genuine danger, with its associated imaginaries, arguments, desires, and policies, promotes novel forms of environmental citizenships that align with the CCP's governance goals. By using cooperative governance mechanisms at the grassroots level, SGOs create a panoply of depoliticized local projects and actions that often revolve around measures of consensus, agreement or making individuals more accountable for taking care of the environment. This is mostly due to their expertise, deep understanding and knowledge of citizens' concerns and ability to encourage volunteer participation and shared identity among residents. SGOs end up reinforcing Xi Jinping's discourses on creating a "beautiful China", absolving the Party of environment responsibility, consequently, lose the power of advocacy from the ground up.

More broadly, allowing more participation and involvement of new non-state actors in governance may lead to counter-discourses and protests. Institutionalizing the capacity for self-governance and circumscribing public debate through depoliticized imaginaries of the "global city", however, has considerable benefits for the CCP: (1) it creates new resources through the delivery of social welfare services, diminishing pressure and responsibility for the party-state; (2) it enhances social control because different parties surveil each other (the local state surveils the SGOs and vice-versa); (3) bottom-up participation strengthens the regime's legitimacy among the population because they feel their voice is being heard; and (4) the dissenting voices become inaudible against the backdrop of a majority "green" consensus. A double strategy of coercive and "participatory" mechanisms is key to explaining the CCP's capacity for resilience on the environmental crisis and the efforts of Chinese leaders to reinforce their presence at the grassroots level.

A Chinese "green" model of sustainability?

China is increasingly advancing itself as the "powerbroker"[3] in international environmental talks and, consequently, bringing into question the idea that sustainable development and democracy do overlap. Furthermore, the CCP is expanding a particular narrative on "environmentalism with Chinese characteristics" which, according to the CCP, has completely unique characteristics as compared to western societies. From avoiding the problem to taking a proactive stance on environmental issues, China's strategy on environmental issues has considerably transformed in recent years, in particular since 2008 (Engels 2018). But how unique is the "ecological civilization" discourse advanced by the CCP?

To respond to this question, we urgently need to analyze the specificity of China's environmental governance mechanisms in the context of its authoritarian political system. Of course, this is a complex task. As Kevin Lo (2015, 152) stresses: "studying the nature of environmental governance as a complex process requires a thorough understanding of not just national policy, but also local politics and the ways the two are connected". By pointing out the contentious confrontations of interests and ideologies lying behind China's electric power system, the findings outlined in Coraline Goron's dissertation (2017) provided evidence for this complexity. Despite central directives, the combination of political and financial constraints means that local officials are determined to pursue short-term GDP growth rather than long-term economic and environmental sustainability (Goron 2018, 329). Similarly, Eaton and Kostka (2014) argue that local officials' short-term horizons offer little optimism about the development of non-democratic approaches to environmental policy.

Considering the CCP's decentralized system and the fight against corrupt elites (Gilley 2012), one key question as outlined by Mark Beeson (2018) is whether the state is able to maintain an authoritarian model of environmentalism. For Beeson (2010), *environmental authoritarianism* as a discourse of environmental governance is normally defined by two aspects: (1) a policy process dominated by an autonomous state which is non-participatory—absent from public consultation, grassroots activism, free press, lobbying, or self-interests; and (2) environmental outcomes are pursued by restricting individual liberty through the use of a regulated-based policy.

As I showed in Chapter 3, however, and particularly since Xi Jinping assumed power in 2013, a new set of top-down policies were implemented to respond to several of those issues. I specifically showed that many measures were designed to (re)centralize decision-making and strengthen central control:

> In this phase of tightening up, the power balance between central and local governments has been tipped decisively in the centre's favour as Xi has removed powers and discretion from local governments, introduced new monitoring and sanctioning practices, and signalled a zero-tolerance approach to non-compliance with central directives by sending thousands of local officials to prison.
>
> (Kostka and Nahm, 2017, 568)

Yet, as this book has shown, the CCP's environmental policies go beyond top-down centralization measures, and I would therefore argue that Beeson's definition of *environmental authoritarianism* needs to be adjusted. Indeed, environmental governance is not fixed or standardized but is continuously strengthened to adapt to changing strategic, economic, social, or environmental realities. The *China "Green" Consensus* claims that, although China hopes to be seen as different from the liberal world, its political discourse on environmental issues falls into the post-political condition observed by many foreign scholars in various other contexts.

My empirical analysis shows that the CCP's rhetoric of "ecological civilization", strong commitment to climate strategies, and effort to green its urban landscape do not represent solely environmental goals, but provide a way to curate a consensus pushed towards depoliticizing environmental politics and enhance China's (and, of course, the Party's) image at both national and international levels. One of my claims is that the party-state has managed to selectively incorporate "environmental goals" while positioning itself as a prime actor in environmental protection. To do so, the regime developed three important strategies which complement one another: (1) (re)centralizing environmental governance efforts; (2) creating a consensus around environmental protection; and (3) institutionalizing grassroots movements. While scholars expect to find some of those strategies in authoritarian settings, others, such as spaces for dialogue and compromise between the government and the society, seem to be more characteristic of democracies. Below, I explore each of these strategies and argue that they have become a turning point for the CCP's authoritarian stability.

(Re)centralizing environmental oversight—but not too much

China's geography and demography differ sharply from those of Singapore, which, according to several authors, follows an environmental authoritarian model (Ortmann 2009, 2016; Han 2017) in a climate of post-politics (Neo 2021) (see Chapter 6). This partly explains why, in China, leaders implement and adapt environmental governance campaigns at a city or provincial level.

In the previous chapters, I have shown that, even though Shanghai follows national goals, policymakers make environmental governance strategies for the metropolis to follow their objectives and needs. As the Chinese proverb says: "Heaven is high, and the emperor is far away (*tian gao huangdi yuan* 天高皇帝远)". It is mainly the city's aspiration to continue being recognized as China's "greatest metropolis" that led the city to become particularly committed to creating a sustainable city.

Yet, for this to happen, Xi's strong leadership and ongoing efforts to (re) centralize power need to coexist with the antagonistic need for space to maneuver at the provincial level. According to Grano (2016), the central environmental politics are highly influenced by the process of implementing these policies. In Chapter 3, I showed that several actors and interest groups cross paths within environmental politics, opening a space for a central–local blame-shifting game to emerge. Combined with the context of controlled media, the central government strategically uses this space to distance itself from any blame for environmental degradation. For Ran (2017), "blame politics" acts as a protective shell for the central leadership to safeguard itself from criticism.

Xi's administration has developed new bureaucratic reforms, environmental laws, and new enforcement mechanisms to better centralize and unify the CCP's leadership (e.g., lifetime accountability system for cadres). Simultaneously, a certain level of decentralization is maintained to make sure central leadership is not to be blamed. While the central authorities strongly advanced the CCP's "ecological civilization" rhetoric, the responsibilities and actions on environmental politics are (mostly) made at the local level. I believe this represents the first of three characteristics that explain the party-state's ability to "reload" its authoritarian rule and fight the danger of an "environmental public awakening". This first characteristic points to the powerful, technocratic, and controlling side of the central government and its coercive strategies. As I continue to analyze below, however, the CCP uses a variety of other "softer" persuasive strategies that should not be forgotten.

Constructing a new political framing

To respond to growing environmental concerns, the Party leaders have mounted an aggressive messaging effort to ensure the public knows they are taking care of protecting the environment. As Gilley (2012) points out, a general lack of awareness among the public rendered China's *environmental authoritarianism* less effective. Several residents (during informal interviews during fieldwork) argued that the one responsible for taking care of the treatment of waste should be the government, not them. But this is quite problematic for the one-party state. As stressed by Ahlers and Shen (2018), the situation became even more problematic when the battle against air pollution gained momentum at a national level. The public reaction following the release of *Under the Dome* was unprecedented. While previous environmental movements focused on specific issues or only reached small portions of the population (e.g., protecting the Tibetan antelope, demonstrations against chemical plants), air pollution directly affected the lives of all strands of the society in big, developed cities, making it far harder to silence protests. As a matter of comparison, the Nujiang River movements against hydropower stations—recognized for their reach across several nations—could not assemble such a powerful response from the public:

> Seen as a health risk that can affect each individual member of society, air pollution now transgresses the usual boundaries between rulers and the ruled, policymakers and policy addresses. These societal irritations can sometimes bear strange fruit: "smog" (*wùmái* 雾霾) and "PM2.5" have become "winged words" among the young and old, and everyday activities are often planned according to what mobile phone apps declare to be "good air" or "bad air" days. Although air pollution has not yet triggered large-scale public unrest or inspired a nationwide environmental movement in China, local communities have become increasingly aware of their rights and now demand increased environmental and health protection.
>
> (Ahlers and Shen, 2018, 303)

Yet unlike issues that can be blamed on corrupt officials or a malicious industry, air pollution implies the direct participation of the public in its resolution. The party-state is aware of this and has been using the force of the state's propaganda to promote this point of view. In the field, particularly during my later visits in 2017–2018, environmental propaganda posters gained prominence in the cityscape, some stating that "Protecting the environment is protecting yourself (*yi baohu huanjing, jiushi baohu ziji* 一保护环境　就是保护自己)" or that "We need to save energy, otherwise we are the ones who will end up 'buried' (*jieyue nengyuan, fouze zuihou bei maimo de haishi women ziji* 节约能源　否则最后被埋没的还是我们自己)" (see Figure 6.1).

Air pollution also became a major concern because it affects everyone including developed urban areas and, in particular, the group that appears to be China's greatest success, its well-educated middle-class. This new middle-class is characterized by being more individualistic and empowered, but also keener to express its discontentment. As such, for the first time in China's long history, the CCP needs to deal with a strong middle-class sitting at the intersection of its ruling elite and the masses. To respond to the potential threat of the rising expectations of this body of the population, the CCP is carefully designing its "green" ambitions to match their needs. More specifically, as green ideas become a consensus among the Chinese population, the state machinery is deliberately instrumentalizing "urban sustainability" in a form that advances certain interests and visions of the "sustainable" city. As Strittmatter (2019, 140) argues: "The Party has for a long time been making these middle classes into the country's most satisfied citizens, and therefore its greatest allies".

Within this wider strategy, the government is deliberately turning the public into a central actor in environmental governance. Although powers of delegation and decision-making stay at the governmental level, there is a push towards encouraging responses on an individual level which prevents collective opposition to projects. Thus, as this book asserted, the background of environmental politics cannot be grasped by limiting our study to the macro-level analysis of China's policies and regulations. Scholarship needs strong empirical evidence to assess the various approaches developed by the party-state to resist the environmental crisis and renew its governance system. In Shanghai, advancing the individual as a responsible actor in protecting its environment has become a central focus of the municipal government. In other words, the aim is to create an atmosphere where "Everybody, as an individual, can work for environmental protection". Not only is public engagement showcased as the answer to tackling waste or air pollution issues, but citizens are also criticized when they are not engaged in protecting their environment. Because a broad consensus on environmental protection has evolved within Chinese society, it becomes increasingly difficult to position oneself against the Party line.

Thus, while it seems that the CCP's environmental authoritarian approach is aligning itself to the many concerns of its population, I contend that the "greening" of the state rather signals an expansion of the boundaries of the CCP (through its narrative) into the daily lives of Chinese citizens. Of course, strengthening the power of selected SGOs to involve the public in a concerted and controlled way is a significant part of this wider strategy.

(Re)aligning public participation in a coordinated framework

Following the ecological implosion in the 1980s, the Chinese government became aware of the social uprisings that an environmental crisis could cause. In reaction, they began introducing environmental laws, but also welcoming assistance from non-state actors and international NGOs. In 1994, the National People's Congress passed the Rules for Registering Social Organizations, which for the first time granted legal status to independent organizations. Since the early 1990s, China's leaders were conscious of the role that non-state and public-led organizations could play in addressing a broad range of emerging social and environmental issues.

The historical evolution of the regulatory system governing SGOs has been complex and marked by periods both of opening and closure (see Chapter 3). I had previously explained how, as part of the semi-privatization of governance, many of the state's traditional roles and responsibilities were outsourced to various types of SGOs (Gao and Tyson 2017). For a long time, by limiting and restricting their development, the CCP has managed to maintain a certain level of control over groups seen as undesirable. Still, the "informality" of the sector enabled several SGOs (either accepted and/or not accepted by the regime) to slowly increase their political space (Fulda 2017).

Yet, since Xi Jinping came to power, researchers seem to have reached a new consensus: the attitude towards foreign and non-state actors has changed.[4] Although, as demonstrated in Chapters 4 and 5, SGOs are still able to develop through various means, their activities are being increasingly incorporated into a consensus-oriented decision-making process. Some scholars describe these emerging forms of cross-boundary governance as "collaborative governance" (Ansell and Gash 2008; Jing 2015; Teets and Jagusztyn 2015; Jing and Hu 2017), more specifically for Jing "the sharing of power and discretion within and across the public, nonprofit, and private sectors for public purposes" (2015). Although I strongly agree that exchanges between local governments and SGOs are growing (see Chapter 4), I disagree with Jing's description of "collaborative governance" as a process of "sharing power". Rather, I would suggest that there is a "delegation of power" from the government to certain types of SGOs. Therefore, I argue that the term "cooperative" better captures current state–non-state dynamics. In this book, I showed that the local government has developed certain tools to shape or eliminate SGOs soon after they are set up. But there is not a "sharing of power": rather, SGOs are chosen selectively because they are useful to the regime and their activity can be quickly absorbed and institutionalized to best fit government priorities, the case of ZeroWaste (see Chapter 4) being a case in point here.

Although these "means" look less authoritarian in nature compared to harsher top-down enforcement mechanisms, the fact is that the state has learned how to entice the non-profit sector into a cooperative relationship. I want to reinforce the idea that the organizations on which I based this book were not GONGOs (state-owned NGOs) but rather typical grassroots organizations similar to those found in European democratic contexts, meaning that these SGOs' leaders created them from the bottom-up. Similar to many SGOs in the west, some end-up cooperating with the state for various reasons. Even though their activities are facilitated and co-opted by the state, the organizations observed here (see Appendix B and Appendix C) were not created with the ambition of "serving" the state. Still, the state's sponsorship, or tolerance of the work of these SGOs is by no means aimed at "sharing power" or decision-making. Their relationship might seem the very opposite of the fundamentals of authoritarianism because the cooperation of SGOs with the local government greatly resembles the experience of various SGOs in European countries, but the comparison also depends on the lens through which we observe such dynamics. In the field, many members of the public that I talked to assumed that SGOs were part of the government. Of course, this association with a governmental identity did not please ZeroWaste's staff, for instance. On several occasions (during interviews and volunteering activities) staff acknowledged their frustration, and expressed their understanding that their close ties to the government might confuse Chinese citizens. The directors of ZeroWaste and BlueOcean (registered as NGOs) both shared a negative view of GONGOs (Interviews April, May 2016). As one of them argued, GONGOs are not efficient because they are artificially constructed, they don't establish close relations with citizens, and their main aim is to please those further up in the party hierarchy.

Therefore, according to the empirical evidence gathered from the field, I argue that the party-state is strategically orienting grassroots and locally based SGOs to cope with the various and increasing challenges they face in urban areas. Environmentalism is a flourishing area to engage such actors, in particular, because Chinese leaders came to recognize the potential for mobilization and local knowledge capacity of such organizations. So, we can conclude that the CCP's approach to governance shares several characteristics that, at first glance, may seem to have more in common with democratic settings: promoting new forms of "good citizenship", central to enhancing the city's competitiveness and "sustainable" development. I argue that the approach is particularly successful because the government thereby manages to meet the goals of SGOs while harnessing the different organizations' ability to work on the issues at hand. As such, *China's Green Consensus* emphasizes a counterintuitive link between authoritarian resilience, a growing number of grassroots SGOs and, of course, public participation.

Consensus politics: in search of alternative socio-environmental imaginaries

Environmental issues have the power to bring different voices together and influence the possible future evolution of different societal structures. People can internalize environmental issues and give them specific content, and so a certain political space of plurality can emerge, where conflict, contingencies, and power become visible and contestable (Kenis and Lievens 2014). An anti-dam movement, for example, can either advance the rights of the people who might be displaced, the flora and fauna that will be put in danger, or argue in favor of the "green" energy it will produce. As this book has suggested, the environment can also become an ideology within which a particular understanding of "nature" forecloses expression of alternative views and political debate. I have particularly questioned the view of the CCP as monolithic and brought attention to the ways in which political leaders are grooming "green" narratives to establish a frame of reference for consensus building and, thus, displacing disagreements and conflicts. I draw on Actor-Network Theory (ANT), showing how this perspective is useful to analyze how urban norms, projections, and structures unfold. Despite a few limitations, this approach offers several advantages that help appreciate the complexity of reality and the active role of a myriad of actors (human and non-human) in shaping China's "green" consensus.

In the same way, I see post-politics as a promising conceptual tool to further assess practices of depoliticization in authoritarian regimes, in particular, for enhancing our understanding of how environmental politics are used—or can be used—to defuse dissent and create hegemonic consensus. As I have tried to show, such conceptual tools help distinguish tactics of depoliticization from types of power that are exerted through less obvious repression in an authoritarian context. Yet, perhaps even more important at this stage, scholars could use ANT in future research to expose acts of resistance. I would therefore like to end this (mostly) pessimistic analysis of China's current environmental politics on a positive note. Although access to the field is likely to become more difficult in the coming years, scholars must insist on debunking bottom-up visions of sustainability put forward by parties who are trying to (re)politicize the conversation. In particular, we must continue to shine a light on movements, visions, and acts of undiscipline focused on reconnecting environmental, social, and economic issues.

Notes

1 Li Dongliang (2020) "*Dazao gong jian gong zhi gongxiang shehui zhili geju* 打造共建共治共享社会治理格局", Chinese People's Political Consultative Conference Shanghai Committee, available at http://www.shszx.gov.cn/node2/node5368/node5382/node5400/u1ai106294.html (accessed 12 October 2021).
2 Here I take inspiration from Swyngedouw (2013), who defines the "consensual scripting" of climate change as clear and present danger, and its associated imaginaries, arguments and policies arguably reflects a particular process of de-politicization.
3 This term was taken from L. Hook and L. Hornby (November 2018) "China emerges as powerbroker in global climate talks", *Financial Times,* available at https://www.ft.com/content/7c1f16f8-e7ec-11e8-8a85-04b8afea6ea3 (accessed 27 May 2019).
4 Thomas E. Kellogg (24 January 2017) "Xi's Davos Speech: Is China the New Champion for the Liberal International Order?", *The Diplomat.* Retrieved from: https://thediplomat.com/2017/01/xis-davos-speech-is-china-the-new-champion-for-the-liberal-international-order/ (accessed 11 November 2018).

References

Ahlers, Anna L., and Yongdong Shen. 2018. "Breathe Easy? Local Nuances of Authoritarian Environmentalism in China's Battle against Air Pollution". *The China Quarterly* 234: 299–319.

Ansell, Chris, and Alison Gash. 2008. "Collaborative Governance in Theory and Practice". *Journal of Public Administration Research and Theory* 18 (4): 543–571.

Beeson, Mark. 2010. "The Coming of Environmental Authoritarianism". *Environmental Politics* 19 (2): 276–294.

Beeson, Mark. 2018. "Coming to Terms with the Authoritarian Alternative: The Implications and Motivations of China's Environmental Policies". *Asia and the Pacific Policy Studies* 5 (1): 34–46.

Eaton, Sarah, and Genia Kostka. 2014. "Authoritarian Environmentalism Undermined? Local Leaders' Time Horizons and Environmental Policy Implementation in China". *The China Quarterly* 218 (1): 359–380.

Engels, Anita. 2018. "Understanding How China Is Championing Climate Change Mitigation". *Palgrave Communications* 4 (1): 101.

Fu, Diana, and Greg Distelhorst. 2018. "Grassroots Participation and Repression under Hu Jintao and Xi Jinping". *China Journal* 79 (1): 100–122.

Fulda, Andreas. 2017. "The Contested Role of Foreign and Domestic Foundations in the PRC: Policies, Positions, Paradigms, Power". *Journal of the British Association for Chinese Studies*7.

Gao, Hong, and Adam Tyson. 2017. "Administrative Reform and the Transfer of Authority to Social Organizations in China". *The China Quarterly* 232: 1050–1069.

Gilley, Bruce. 2012. "Authoritarian Environmentalism and China's Response to Climate Change". *Environmental Politics* 21 (2): 287–307.

Goron, Coraline. 2017. *Climate Revolution or Long March? The Politics of Low-Carbon Transformation in China (1992–2015). The Power Sector as Case Study.* Université Libre de Bruxelles. PhD Dissertation.

Goron, Coraline. 2018. "Ecological Civilisation and the Political Limits of a Chinese Concept of Sustainability". *China Perspectives* 2018 (4): 39–52.

Grano, Simona A. 2016. "China's Changing Environmental Governance: Enforcement, Compliance and Conflict Resolution Mechanisms for Public Participation". *China Information* 30 (2): 129–142.

Han, Heejin. 2017. "Singapore, a Garden City: Authoritarian Environmentalism in a Developmental State". *Journal of Environment and Development* 26 (1): 3–24.

Jing, Yijia. 2015. *The Road to Collaborative Governance in China. The Road to Collaborative Governance in China.* Palgrave Macmillan.

Jing, Yijia, and Yefei Hu. 2017. "From Service Contracting To Collaborative Governance: Evolution of Government-Nonprofit Relations". *Public Administration and Development* 37 (3): 191–202.

Kenis, Anneleen, and Matthias Lievens. 2014. "Searching for 'the Political' in Environmental Politics". *Environmental Politics* 23 (4): 531–548.

Kostka, Genia, and Jonas Nahm. 2017. "Central–Local Relations: Recentralization and Environmental Governance in China". *The China Quarterly* 231: 567–582.

Lo, Kevin. 2015. "How Authoritarian Is the Environmental Governance of China?" *Environmental Science and Policy* 54: 152–159.

Mouffe, Chantal. 2016. *L'illusion Du Consensus.* Albin Michel.

Neo, Harvey. 2021. "The Post-Politics of Environmental Engagement in Singapore". In *Environmental Movements and Politics of the Asian Anthropocene*, edited by Paul Jobin, Ming-sho Ho, and Michael Hsin-huang Hsiao, 109–138. ISEAS—Yusof Ishak Institute.

Ortmann, Stephan. 2016. *Environmental Governance under Authoritarian Rule: Singapore and China.* Southeast Asia Research Centre (SEARC) Working Paper Series, no. 189, 1–26. Southeast Asia Research Centre, Hong Kong. http://www.cityu.edu.hk/searc/Resources/Paper/16100610_189%20-%20WP%20-%20Dr%20Ortmann.pdf.

Ran, Ran. 2017. "Understanding Blame Politics in China's Decentralized System of Environmental Governance: Actors, Strategies and Context". *China Quarterly* 231: 634–661.

Strittmatter, Kai. 2019. *We Have Been Harmonised: Life in China's Surveillance State.* Old Street Publishing.

Swyngedouw, Erik. 2013. "The Non-Political Politics of Climate Change". *ACME: An International Journal for Critical Geographies* 12 (1): 1–8.

Swyngedouw, Erik. 2019. *Promises of the Political: Insurgent Cities in a Post-Political Environment.* MIT Press.

Teets, Jessica C., and Marta Jagusztyn. 2015. "The Evolution of a Collaborative Governance Model: Social Service Outsourcing to Civil Society Organizations in China". In *NGO Governance and Management in China*, 69–88. Routledge.

Appendix A: Semi-structured interviews

To preserve the identity and safety of some of my informants, I erased some of the names and/or any information that could lead to a recognition of their identity. A summary of the informant's features is available below.

First field research

March to May 2016

- Associate Professor in Environmental Education and Education for Sustainability.
  5 April 2016, Minhang, Shanghai.
- Program Manager at WWF China.
  18 April 2016, WWF locals, Hongkou, Shanghai.
- Director of Turtle (see Appendix C).
  19 April 2016, Lujiazui, Shanghai.
- Director of ZeroWaste.
  26 April 2016, ZeroWaste locals, Shanghai.
- Director of BlueOcean.
  16 May 2016, NPI locals, Shanghai.
- General manager at Goat (see Appendix C).
  6 May 2016, Goat locals, Shanghai.

Second field research

September 2016 to June 2017

- Marketing and Communication Manager, Non-profit Incubator (NPI).
  27 October 2016, 724 Cheers Hub, Pudong, Shanghai.
- Professor of Education for Sustainable Development
  6 November 2016, National Taiwan Normal University, Taiwan.
- Co-founder of Pig (see Appendix C).
  21 November 2016, Sandbox, Jing'an, Shanghai.

— Weina, product innovation manager for a French multinational, planned to open a social enterprise shortly after the interview; active volunteer of ZeroWaste and other SGOs mostly focusing on environmental issues.
21 November 2016, Sandbox, Jing'an, Shanghai.
— Song (fictitious name), Founder of Farming.
24 November 2016, Anfu Road, French Concession, Shanghai.
— Professor at Tsinghua University, School of Marxism.
1 December 2016, Tsinghua University, Beijing.
— Chloé Froissart, Director of the Sino-French Centre, Tsinghua University, Beijing.
1 December 2016, Tsinghua University, Beijing.
— Program Manager at NPI.
21 December 2016, 724 Cheers Hub, Pudong, Shanghai.
— Founder of a social enterprise focused on improving the environment through exercise activities and eco-tourist packages (a collaborator of Farming).
28 February 2016, Sandbox, Jing'an, Shanghai.
— Tiffany Pattinson, Eco-Chic designer and social entrepreneur.
24 March 2017, Fresh Start meeting, Kerry Centre, Jing'an, Shanghai.
— Founder and Managing Director of Lion (see Appendix C).
29 March 2017, Xujiahui, Shanghai.
— Communication Manager, SynTao Green Finance, a consultancy providing professional services in green finance and responsible investment in China.
19 April 2017, Pudong, Shanghai.
— Co-founder and Operations Director at Mouse (see Appendix C).
9 May 2017, Jing'an, Shanghai.
— Co-founder and Program advisor at Snake (see Appendix C).
22 June 2017, Fresh Start Rotary meeting, Naked Hub, Nanjing West Road, Shanghai.
— Past-President (2009–2010) of an international service organization in Shanghai China. He is also the organizations' Special Representative to China.
30 June 2017, Shanghai.

Appendix B: Observed registered SGOs

In the field, I met several officially registered SGOs. In the list below, I briefly present the SGOs from which I retrieved most of the empirical data I present in the book (mostly in Chapters 3, 4, and 5). Because of political sensitivity, I anonymized their names. A brief description of their features is available below.

ZeroWaste

ZeroWaste is the case study presented in Chapter 4. Established in 2012, the organization focuses on community waste management and waste reduction. It mainly promotes community practices, training and consultation, community education and public advocacy in Shanghai communities.

BlueOcean

Registered as a civil non-enterprise unit since 2007, BlueOcean is the only registered organization in Shanghai focused on marine environmental protection. I developed several volunteering activities with BlueOcean, ranging from environmental-related classes or beach clean-ups.

Garden

Garden mainly promotes environmental-related projects such as waste segregation, education, and community city vegetables gardens. I particularly observed the implementation of the Bottle Garden Project 瓶子菜园 in several Shanghainese communities.

Student

Student is a student association affiliated with a Chinese University. The organization was introduced to me by my colleagues at ECNU. I accompanied them on several occasions in primary school activities and educational workshops on campus.

Appendix C: Characteristics of the analyzed social enterprises

In the list below, I present the various SGOs (either Chinese or international) which I observed in the field. Similarly to the SGOs presented in Appendix B, I purposefully anonymized. The list below mostly refers to the organizations I used for Chapter 5.

	Sector	*Type*	*Date*	*Staff*	*Funding*	*Members*	*Origin*	*Registration*
Snake	Charity and migrants	NGO/SE	2009	8–11	Chinese and Hong Kong Foundations	Started by foreigners but more and more managed by Chinese	Shanghai	Registered as an NGO in Hong Kong since 2009. The organization started its own social enterprise in 2011 in China.
Whale	Social entrepreneurship and charity platform	NGO/SE	2006	105	Chinese foundations	Chinese	Shanghai	Registered as a Civil Non-Enterprise Unit in China
Turtle	Education and Environment	non-profit/ SE	2009	2–3	Sponsorships from the private sector	Started by foreigners with more and more Chinese participants	Shanghai	Used to be registered as a for-profit company in Hong Kong. Registered as a foreign company in China since 2017
Wolf	Disabled	SE	2010	12	Business model	Chinese	Shanghai	Registered as a Civil Non-Enterprise Unit in China
Tiger	Community platform	SE	2016	5–10	Business model	Chinese with a strong foreign environment	Shanghai	Registered as an enterprise in China
Panda	Gender equality	SE	2016	2–3	Sell merchandise	Chinese	Shanghai	Registered as an enterprise in China

	Sector	*Type*	*Date*	*Staff*	*Funding*	*Members*	*Origin*	*Registration*
Bear	**Ecological products**	SE	2011	10–30	Business Model	Chinese with a strong foreign environment	Shanghai	Registered as an enterprise in China
Koala	Organic and sustainable food	SE	2015	5–10	Business model	Chinese with a strong foreign environment	Shanghai	Registered as an enterprise in China
Mouse	Waste reduction	SE	2016	2–3	Business model	Foreign	Shanghai	Registered as an enterprise in China
Farming	Agriculture	SE	2016	5–10	Community Supported Agriculture	Chinese with a strong foreign environment.	Shanghai	Registered as an enterprise in China since 2016. Since 2018 work as the sister organization of a social enterprise registered in Hong Kong.
Cat	Homeless	SE	2008	2–8	Donations/ Business model	Foreign	Shanghai	Foreign consultancy company in China
Lion	Education	SE	2011	2–3	CSR Business model	Foreign	Shanghai	Registered as an enterprise in China
Pig	Entrepreneurship incubator	SE	2015	2–5	Business model	Foreign	Shanghai	Registered as an enterprise in China.

	Sector	*Type*	*Date*	*Staff*	*Funding*	*Members*	*Origin*	*Registration*
Eagle	Social and cultural events	SE	2015	2–5	Business model	Foreign	Shanghai	Registered as an enterprise in China
Goat	E-waste pollution reduction and poverty relief	SE	2012	5–10	CSR activities	Chinese with a strong foreign environment. Initially started by foreigners.	Shanghai	Registered as a social enterprise in Hong Kong since 2012. Opened a subsidiary operational company in China in 2013.

Index

D

For Product Safety Concerns and Information please contact our EU
representative GPSR@taylorandfrancis.com
Taylor & Francis Verlag GmbH, Kaufingerstraße 24, 80331 München, Germany

www.ingramcontent.com/pod-product-compliance
Lightning Source LLC
Chambersburg PA
CBHW060302220326
41598CB00027B/4207
9781032138831